FROM RELATIVITY TO QUANTUM MECHANICS

Frank W. K. Firk

2019

Copyright © 2019 Frank W K Firk

All rights reserved

ISBN: 9781688314788

Printed in USA KDP, 2019

TABLE OF CONTENTS PAGE

PREFACE	5
1. SOME INVARIANTS IN MATHEMATICS	7
2. NEWTONIAN RELATIVITY AND THE GALILEAN TRANSFORMATION	18
3. EINSTEINIAN RELATIVITY, SPACE-TIME SYMMETRY AND THE LORENTZ TRANSFORMATION	22
4. THE 4–VELOCITY	28
5. THE RELATIVITY OF SIMULTANEITY, LENGTH CONTRACTION AND TIME–DILATION	29
6. RELATIVISTIC COLLISIONS AND THE CONSERVATION OF MOMENTUM	36
7. RELATIVISTIC INELASTIC COLLISIONS	39
8. THE MANDELSTAM VARIABLES	42
9. THE FORMAL STRUCTURE OF LORENTZ TRANSFORMATIONS	45
10. ELECTROMAGNETISM: RELATIVITY IN ACTION	47
11. A RELATIVISTIC THEORY OF FIELDS	55
12. RELATIVITY AND PARTICLE – WAVE DUALITY	64
13. RELATIVITY AND deBROGLIE MATTER-WAVES	70

14. ENERGY AND MOMENTUM OPERATORS 76
    IN QUANTUM MECHANICS:
    THE SCHROEDINGER EQUATION
15. THE FOURIER–RAYLEIGH– deBROGLIE– 82
    HEISENBERG UNCERTAINTY PRINCIPLE
16. THE KLEIN – GORDON EQUATION 93
17. THE DIRAC EQUATION 95
18. DIRAC MATRICES, WAVE FUNCTIONS, 99
    ELECTRONS AND POSITRONS
19. INTRODUCTION TO LIE GROUPS 102
20. THE SPECIAL UNITARY GROUP SU(2) 104
21. LIE'S CONTINUOUS 107
    TRANSFORMATION GROUPS
22. ROTATIONS, ANGULAR MOMENTUM 112
    AND LIE GROUPS
23. NEUTRONS, PROTONS AND 115
    THE YUKAWA POTENTIAL
24. ELECTRON – POSITRON 118
    ANNIHILATION – IN – FLIGHT
25. PROBABILITY, AND THE MAXWELL - 122
    BOLTZMANN EQUATION

PREFACE

It is the tradition to present the two branches of Physics – *Relativity and Quantum Mechanics* – as independent subjects. The subject matter in both branches is extensive and requires years of study to appreciate the material. In this booklet, basic concepts of both topics are presented in an approach that emphasizes the essential connection between the *invariants* of *Einstein's Relativity* and *deBroglie's Particle–Wave Duality*. The Schroedinger equation and Dirac's Relativistic Quantum Mechanics follow naturally. Key ideas that involve particle spin, and matter and anti-matter, emerge in Dirac's formal theory. The final, different topic, is the Maxwell-Boltzmann equation.

!. SOME NVARIANTS IN MATHEMATICS

It is a remarkable fact that very few fundamental laws are required to describe the enormous range of physical phenomena that take place throughout the universe. The study of these fundamental laws is at the heart of Physics. The laws are found to have a mathematical structure; the interplay between Physics and Mathematics is therefore emphasized throughout these essays. For example, Galileo found by observation, and Newton developed within a mathematical framework, the Principle of Relativity: *the laws governing the motions of objects have the same mathematical form in all inertial frames of reference.*

Inertial frames move at constant speed in straight lines with respect to each other – they are non-accelerating. Newton's laws of motion are *invariant* under the so-called Galilean transformation. The discovery of key *invariants* of Nature has been essential for the development of the subject.

Einstein extended the Newtonian Principle of Relativity to include the motions of beams of light and of objects that move at speeds close to the speed of light. This extended principle forms the basis of Special Relativity. Later, Einstein generalized the principle to include accelerating frames of reference. The general principle is known as the Principle of Covariance; it forms the basis of the General Theory of Relativity (a theory of Gravitation).

Geometrical invariants

A non-traditional proof of the most famous theorem of Euclidean Geometry, namely Pythagoras' theorem, is based on the *invariance* of length and angle (and therefore of area) under translations and rotations in space. Let a right-angled triangle with sides a, b, and c, be translated and rotated into the following four positions to form a square of side c:

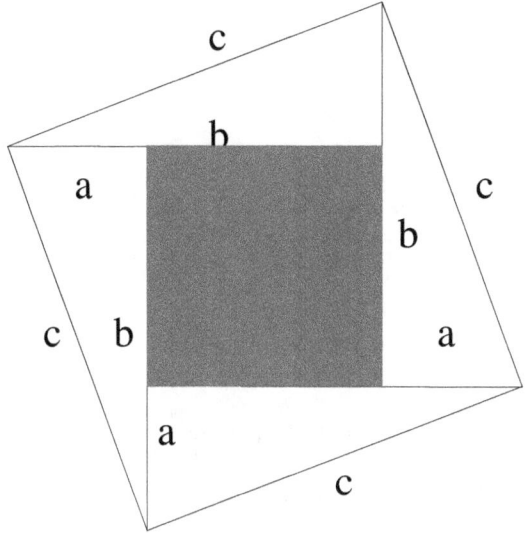

The total area of the square = c^2 = area of four triangles + area of shaded square $(b - a)^2$

If the four right-angled triangles are translated and rotated to form the rectangle:

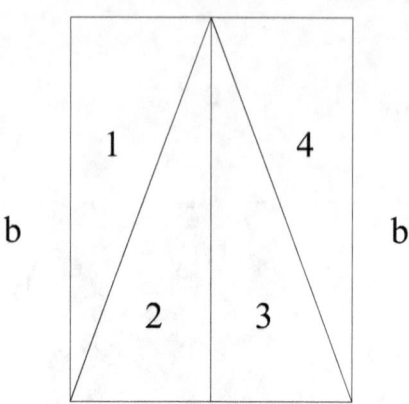

then the area of four triangles = 2ab.

The area of the shaded square area is

$$(b-a)^2 = b^2 - 2ab + a^2$$

Invoking the invariance of length and angle under translations and rotations we have

$$c^2 = 2ab + (b-a)^2$$
$$= a^2 + b^2 \text{ (Pythagoras' theorem)}$$

This key result characterizes the locally flat space in which we live. *It is the only form that is consistent with the invariance of lengths and angles under translations and rotations.*

The Invariant Scalar Product

The *scalar product* is an important invariant in Mathematics and Physics. The invariance properties of the product can best be seen by developing Pythagoras' theorem in a three-dimensional coordinate form. Consider the square of the distance between the points $P[x_1, y_1, z_1]$ and $Q[x_2, y_2, z_2]$ in Cartesian coordinates. We have

$$(PQ)^2 = (x_2 - x_1)^2 + (y_2 - y_1)^2 + (z_2 - z_1)^2$$
$$= x_2^2 - 2x_1x_2 + x_1^2 + y_2^2 - 2y_1y_2 + y_1^2 + z_2^2 - 2z_1z_2 + z_1^2$$
$$= (x_1^2 + y_1^2 + z_1^2) + (x_2^2 + y_2^2 + z_2^2)$$
$$- 2(x_1x_2 + y_1y_2 + z_1z_2)$$
$$= (OP)^2 + (OQ)^2 - 2(x_1x_2 + y_1y_2 + z_1z_2)$$

where O is the origin of the coordinates.

The lengths PQ, OP, OQ, and their squares, are invariants under rotations and therefore the entire right-hand side of this equation is an invariant. *This means that the admixture of the coordinates*

$(x_1x_2 + y_1y_2 + z_1z_2)$ *is an invariant under rotations.* This term has a geometric interpretation: in the triangle OPQ, we have the generalized Pythagorean theorem

$$(PQ)^2 = (OP)^2 + (OQ)^2 - 2OP.OQ \cos\alpha,$$

where α is the angle between OP and OQ, therefore

$$OP \cdot OQ \cos\alpha = x_1x_2 + y_1y_2 + z_1z_2$$

$$\equiv \textit{the scalar product.}$$

An Algebraic Invariant

Although algebraic invariants first appeared in the works of Lagrange and Gauss in connection with the Theory of Numbers, the study of algebraic invariants as an independent branch of Mathematics did not begin until the work of Boole in 1841. Before discussing this work, it will be convenient to introduce matrix versions of real bilinear forms, B, defined by

$$B = \sum_{i=1}^{m} \sum_{j=1}^{n} a_{ij}x_iy_j$$

where

$$\mathbf{x} = [x_1, x_2, \ldots x_m], \text{ an m-vector,}$$

$$\mathbf{y} = [y_1, y_2, \ldots y_n], \text{ an n-vector,}$$

and a_{ij} are real coefficients. The square brackets denote a *column* vector.

In matrix notation, the bilinear form is

$$B = \mathbf{x}^T A \mathbf{y}$$

where

$$A = \begin{bmatrix} a_{11} & \cdots & a_{1n} \\ \cdot & \cdot & \cdot \\ a_{m1} & \cdot & a_{mn} \end{bmatrix}$$

The scalar product of two n-vectors is seen to be a special case of a bilinear form in which $A = I$. If $\mathbf{x} = \mathbf{y}$, the bilinear form becomes a quadratic form $Q = \mathbf{x}^T A \mathbf{x}$.

Invariants of Binary Quadratic Forms

Boole began by considering the properties of the quadratic form

$$Q(x,y) = ax^2 + 2hxy + by^2$$

under a linear transformation of the coordinates

$$x' = Mx \text{ where}$$

$$x = [x,y],$$

$$x' = [x',y'], \text{ and}$$

$$M = \begin{bmatrix} i & j \\ k & l \end{bmatrix}$$

M transforms an orthogonal coordinate system into an oblique coordinate system as shown:

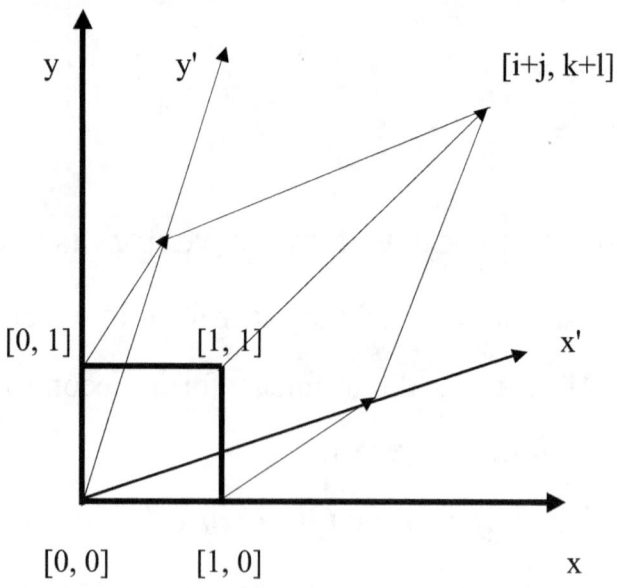

*The transformation of a unit square under **M**.*

The transformation is linear, therefore the new function Q'(x',y') is a binary quadratic:

$$Q'(x',y') = a'x'^2 + 2h'x'y' + b'y'^2.$$

The original function can be written

$$Q(x,y) = \mathbf{x}^T \mathbf{D} \mathbf{x}$$

where

$$\mathbf{D} = \begin{pmatrix} a & h \\ h & b \end{pmatrix}$$

and the determinant of **D** is

$$\det \mathbf{D} = ab - h^2,$$ is the discriminant of Q.

The transformed function can be written

$$Q'(x',y') = \mathbf{x'}^T \mathbf{D'} \mathbf{x'}$$

where

$$\mathbf{D'} = \begin{pmatrix} a' & h' \\ h' & b' \end{pmatrix}$$

and
$$\det \mathbf{D}' = a'b' - h'^2, \text{ the discriminant of } Q'.$$
Now,
$$Q'(x',y') = (\mathbf{M}\mathbf{x})^T \mathbf{D}' \mathbf{M}\mathbf{x}$$
$$= \mathbf{x}^T \mathbf{M}^T \mathbf{D}' \mathbf{M}\mathbf{x}$$
and this is equal to $Q(x,y)$ if
$$\mathbf{M}^T \mathbf{D}' \mathbf{M} = \mathbf{D}.$$
The invariance of the form $Q(x,y)$ under the coordinate transformation $\mathbf{M}$ therefore leads to the relation
$$(\det \mathbf{M})^2 \det \mathbf{D}' = \det \mathbf{D}$$
because
$$\det \mathbf{M}^T = \det \mathbf{M}.$$
The explicit form of this equation involving determinants is
$$(il - jk)^2 (a'b' - h'^2) = (ab - h^2).$$
The discriminant $(ab - h^2)$ of Q is said to be an *invariant* of the transformation because it is equal to the discriminant $(a'b' - h'^2)$ of Q', apart from a factor $(il - jk)^2$ that depends on the

transformation itself, and not on the arguments a, b, h of the function.

We now come to the fundamental questions involving possible invariants associated with physical quantities such as space–time, energy and momentum, and frequency wave–number.

2. NEWTONIAN RELATIVITY AND THE GALILEAN TRANSFORMATION

Events – descriptions of where and when things happen – are not constructs of the mind; they are essential ingredients of the physical world. In general, an event can be described by a (column) four-vector E[t, x, y, z] in a space–time geometry where t is the time, and x, y, z are the Cartesian (spatial) coordinates, referred to an arbitrarily chosen origin. (Other coordinate systems can be used, if more convenient). **

** To be consistent with matrix multiplication *square brackets around numbers* are used to denote a *column of the numbers*: $[1, 2] = \begin{pmatrix} 1 \\ 2 \end{pmatrix}$ eg: if A[a, b] , B[c, d],

and $C = \begin{pmatrix} e & f \\ g & h \end{pmatrix}$,

and A = CB then $\begin{pmatrix} a \\ b \end{pmatrix} = \begin{pmatrix} e & f \\ g & h \end{pmatrix} \begin{pmatrix} c \\ d \end{pmatrix}$ so that

a = ec + fd and b = gc + hd

Let an event E[t, x], recorded by an observer O at the origin of an x-axis, be recorded as the event E´[t´, x´] by a second observer O´, moving at constant speed V along the x-axis. We suppose that their clocks are synchronized at t = t´ = 0 when they coincide at a common origin, x = x´ = 0. At time t, we write the plausible equations

$$t' = t$$

and

$$x' = x - Vt$$

where Vt is the distance traveled by O´ in a time t. These equations can be written

$$E' = GE$$

where

$$G = \begin{pmatrix} 1 & 0 \\ -V & 1 \end{pmatrix}.$$

G is the *operator of the Galilean transformation*. The inverse equations are

and
$$t = t'$$
$$x = x' + Vt'$$
or
$$E = G^{-1}E'$$

where G^{-1} is the inverse Galilean operator. (It undoes the effect of G).

If we multiply t and t′ by the constants k and k′, respectively, where k and k′ have dimensionsof velocity then all terms have dimensions of length.

In space-space, we have the Pythagorean form $x^2 + y^2 = r^2$ (an invariant under rotations). We are therefore led to ask the question: is $(kt)^2 + x^2$ an invariant under G in space-time? Direct calculation gives

$$(kt)^2 + x^2 = (k't')^2 + x'^2 + 2Vx't' + V^2t'^2$$
$$= (k't')^2 + x'^2 \text{ only if } V = 0.$$

We see that Galilean space-time does not leave the sum of squares invariant. Note, however, the key role of acceleration in Galilean-Newtonian physics:

The velocities **v** and **v′** of the events according to O and O′ are obtained by differentiating

$$x' = -Vt + x$$ with respect to time, giving

$$\mathbf{v'} = -\mathbf{V} + \mathbf{v},$$ a result that agrees with observations.

Differentiating **v′** with respect to time gives

$$d\mathbf{v'}/dt' = \mathbf{a'} = d\mathbf{v}/dt = \mathbf{a}$$

where **a** and **a′** are the accelerations in the two frames of reference. The *classical acceleration is an invariant under the Galilean transformation.* If the relationship $\mathbf{v'} = \mathbf{v} - \mathbf{V}$ is used to describe the motion of a pulse of light, moving in empty space at $v = c \cong 3 \times 10^8$ m/s, it does not fit the facts. For example, if V is 0.5c, we expect to obtain v′ = 0.5c, whereas, it is found that v′ = c. In all cases studied, v′ = c for all values of V.

3. EINSTEINIAN RELATIVITY, SPACE–TIME SYMMETRY AND THE LORENTZ TRANSFORMATION

It was Einstein, above all others, who advanced our understanding of the nature of space-time and relative motion. He made use of a symmetry argument to find the changes that must be made to the Galilean transformation if it is to account for the relative motion of rapidly moving objects and beams of light. Einstein recognized an inconsistency in the Galilean-Newtonian equations, based as they are, on everyday experience. The discussion will be limited to non-accelerating, or so called inertial, frames.

We have seen that the classical equations relating the events E and E′ are E′ = GE, and the inverse E = G^{-1}E′ where

$$G = \begin{pmatrix} 1 & 0 \\ -V & 1 \end{pmatrix} \text{ and } G^{-1} = \begin{pmatrix} 1 & 0 \\ V & 1 \end{pmatrix}$$

These equations are connected by the substitution V ↔ –V; this is an algebraic statement of the

Newtonian *principle of relativity*. Einstein incorporated this principle in his theory. He also retained the linearity of the classical equations in the absence of any evidence to the contrary. (Equispaced intervals of time and distance in one inertial frame remain equispaced in any other inertial frame). He symmetrized the space-time equations as follows:

$$\begin{pmatrix} t' \\ x' \end{pmatrix} = \begin{pmatrix} 1 & -V \\ -V & 1 \end{pmatrix} \begin{pmatrix} t \\ x \end{pmatrix}.$$

Note, however, the inconsistency in the dimensions of the time-equation that has now been introduced:

$$t' = t - Vx.$$

The term Vx has dimensions of $[L]^2/[T]$, and not $[T]$. This can be corrected by introducing the invariant speed of light, c — a postulate in Einstein's theory that is consistent with the result of the Michelson-Morley experiment:

$$ct' = ct - Vx/c$$

so that all terms now have dimensions of length.

Einstein went further, and introduced a dimensionless quantity γ instead of the scaling factor of unity that appears in the Galilean equations of space-time. This factor must be consistent with all observations. The equations then become

$$ct' = \gamma ct - \beta\gamma x$$
$$x' = -\beta\gamma ct + \gamma x, \text{ where } \beta = V/c.$$

These can be written

$$E' = LE,$$

where

$$L = \begin{pmatrix} \gamma & -\beta\gamma \\ -\beta\gamma & \gamma \end{pmatrix}$$

and

$$E = [ct, x].$$

L *is the operator of the Lorentz transformation.*

The inverse equation is

$$E = L^{-1}E'$$

where L^{-1} *is the inverse Lorentz transformation,* obtained from L by changing $\beta \to -\beta$ ($V \to -V$); it has the effect of undoing the transformation **L**.

Carrying out the matrix multiplications, and equating elements gives

$$\gamma^2 - \beta^2\gamma^2 = 1$$

therefore,

$$\gamma = 1/\sqrt{(1 - \beta^2)} \text{ (taking the positive root)}$$

As $V \to 0$, $\beta \to 0$ and therefore $\gamma \to 1$; this represents the classical limit in which the Galilean transformation is, for all practical purposes, valid. In particular, time and space intervals have the same measured values in all Galilean frames of reference, and acceleration is the single fundamental invariant.

The invariant interval: contravariant and covariant vectors

Previously, it was shown that the space-time of Galileo and Newton is not Pythagorean under G. We now ask the question: is Einsteinian space-time Pythagorean under L ? Direct calculation leads to

$$(ct)^2 + x^2 = \gamma^2(1 + \beta^2)(ct')^2 + 4\beta\gamma^2 x' ct'$$
$$+ \gamma^2(1 + \beta^2)x'^2$$
$$\neq (ct')^2 + x'^2 \text{ if } \beta > 0.$$

Note, however, that the *difference of squares is an invariant*:

$$(ct)^2 - x^2 = (ct')^2 - x'^2$$

because

$$\gamma^2(1 - \beta^2) = 1.$$

Space-time is said to be *pseudo-Euclidean*. The negative sign that characterizes Lorentz invariance can be included in the theory in a general way as follows. We introduce two kinds of 4-vectors

$$x^\mu = [x^0, x^1, x^2, x^3], \text{ a contravariant}$$

vector,

and

$$x_\mu = [x_0, x_1, x_2, x_3], \text{ a covariant vector,}$$

where

$$x_\mu = [x^0, -x^1, -x^2, -x^3].$$

The scalar (or inner) product of the vectors is defined as

$$x^{\mu T} x_\mu = (x^0, x^1, x^2, x^3)[x^0, -x^1, -x^2, -x^3],$$

$$\phantom{x^{\mu T} x_\mu = (x^0, x^1, x^2, x^3)}\uparrow \uparrow$$

$$\phantom{x^{\mu T} x_\mu = (x^0, x^1, x^2, x)}\text{row} \text{column}$$

$$= (x^0)^2 - ((x^1)^2 + (x^2)^2 + (x^3)^2).$$

The superscript T is usually omitted in writing the invariant; it is implied in the form $x^\mu x_\mu$.

The event 4-vector is

$$E^\mu = [ct, x, y, z],$$ and the covariant form is

$$E_\mu = [ct, -x, -y, -z]$$

so that the invariant scalar product is

$$E^\mu E_\mu = (ct)^2 - (x^2 + y^2 + z^2).$$

A general Lorentz 4-vector x^μ transforms as follows

$$x'^\mu = Lx^\mu$$

where

$$L = \begin{pmatrix} \gamma & -\beta\gamma & 0 & 0 \\ -\beta\gamma & \gamma & 0 & 0 \\ 0 & 0 & 1 & 0 \\ 0 & 0 & 0 & 1 \end{pmatrix}$$

the operator of the Lorentz transformation if the motion of O′ is along the x-axis of O's frame of reference, and $t = t' = 0$ at $x = x' = 0$.

4. THE 4-VELOCITY

A differential time interval, dt, cannot be used in a Lorentz-invariant way in kinematics. We must use the proper time differential interval, dτ, defined by

$$(cdt)^2 - dx^2 = (cdt')^2 - dx'^2 \equiv (cd\tau)^2.$$

The Newtonian 3-velocity is

$$v_N = [dx/dt, dy/dt, dz/dt],$$

and this must be replaced by the 4-velocity

$$V^\mu = [d(ct)/d\tau, dx/d\tau, dy/d\tau, dz/d\tau]$$

$$= [d(ct)/dt, dx/dt, dy/dt, dz/dt](dt/d\tau)$$
$$= [\gamma c, \gamma \mathbf{v}_N].$$

The scalar product is then
$$V^\mu V_\mu = (\gamma c)^2 - (\gamma \mathbf{v}_N)^2 = (\gamma c)^2 (1 - (\mathbf{v}_N/c)^2) = c^2.$$

The magnitude of the 4-velocity is c.

5. THE RELATIVITY OF SIMULTANEITY: LENGTH CONTRACTION AND TIME DILATION

In order to record the time and place of a sequence of events in a particular inertial reference frame, it is necessary to introduce an infinite set of adjacent "observers", located throughout the entire space. Each observer, at a known, fixed position in the reference frame, carries a clock to record the time and the characteristic property of every event in his immediate neighborhood. The observers are not concerned with non-local events. The clocks carried by the observers are synchronized — they all read the same time throughout the reference frame. The process of synchronization is discussed later. It is the job of the chief observer to collect the information concerning the time, place, and characteristic feature of the events recorded by all observers, and to construct the world line (a path in space-time), associated with a particular characteristic feature (the type of particle, for example).

Consider two sources of light, 1 and 2, and a point M midway between them. Let E_1 denote the event "flash of light leaves 1", and E_2 denote the event "flash of light leaves 2". The events E_1 and E_2 are simultaneous if the flashes of light from 1 and 2 reach M at the same time. The fact that the speed of light in free space is independent of the speed of the source means that simultaneity is relative.

The clocks of all the observers in a reference frame are synchronized by correcting them for the speed of light as follows:

Consider a set of clocks located at x_0, x_1, x_2, x_3, ... along the x-axis of a reference frame. Let x_0 be the chief's clock, and let a flash of light be sent from the clock at x_0 when it is reading t_0 (12 noon, say). At the instant that the light signal reaches the clock at x_1, it is set to read $t_0 + (x_1/c)$; at the instant that the light signal reaches the clock at x_2, it is set to read $t_0 + (x_2/c)$, and so on for every clock along the x-axis. All clocks in the reference frame then "read the same time" — they are synchronized. From the viewpoint of all other inertial observers, in their own reference frames, the set of clocks, sychronized using the above procedure, appears to be unsychronized. It is the lack of symmetry in the sychronization of clocks in different reference frames that leads to two non-intuitive results namely, length contraction and time dilation.

inertial reference frame S´. Because it is at rest, it does not matter when its end-points x_1'

and x_2' are measured to give the rest-, or proper-length of the rod,
$$L_0' = x_2' - x_1'.$$
Consider the same rod observed in an inertial reference frame S that is moving with constant velocity $-V$ with its x-axis parallel to the x'-axis. We wish to determine the length of the moving rod; we require the length $L = x_2 - x_1$ according to the observers in S. This means that the observers in S must measure x_1 and x_2 at the same time in their reference frame. The events in the two reference frames S, and S' are related by the spatial part of the Lorentz transformation:
$$x' = -\beta\gamma ct + \gamma x$$
and therefore
$$x_2' - x_1' = -\beta\gamma c(t_2 - t_1) + \gamma(x_2 - x_1).$$
where
$$\beta = V/c \text{ and } \gamma = 1/\sqrt{(1 - \beta^2)}.$$

Since we require the length $(x_2 - x_1)$ in S to be measured at the same time in S, we must have $t_2 - t_1 = 0$, and therefore

$$L_0' = x_2' - x_1' = \gamma(x_2 - x_1),$$

or

$$L_0' \text{(at rest)} = \gamma L \text{ (moving)}.$$

The length of a moving rod, L, is therefore less than the length of the same rod measured at rest, L_0 because $\gamma > 1$.

Consider a clock at rest at the origin of an inertial frame S′, and a set of synchronized clocks at $x_0, x_1, x_2, ...$ on the x-axis of another inertial frame S. Let S′ move at constant speed V relative to S, along the common x -, x′- axis. Let the clocks at x_o, and x_o' be sychronized to read t_0, and t_0' at the instant that they coincide in space. A proper time interval is defined to be the time between two events measured in an inertial frame in which the two events occur at the same place. The time part of the Lorentz transformation can be used to relate an interval of time measured on

the single clock in the S′ frame, and the same interval of time measured on the set of synchronized clocks at rest in the S frame. We have
$$ct = \gamma ct' + \beta\gamma x'$$
or
$$c(t_2 - t_1) = \gamma c(t_2' - t_1') + \beta\gamma(x_2' - x_1').$$
There is no separation between a single clock and itself: $x_2' - x_1' = 0$, so that
$$c(t_2 - t_1)(\text{moving}) = \gamma c(t_2' - t_1')(\text{at rest}) \ (\gamma > 1)$$
A moving clock runs more slowly than a clock at rest.

The general 2 × 2 matrix operator transforms rectangular coordinates into oblique coordinates. The Lorentz transformation is a special case of the 2 × 2 matrices, and therefore its effect is to transform rectangular space-time coordinates into oblique space-time coordinates.

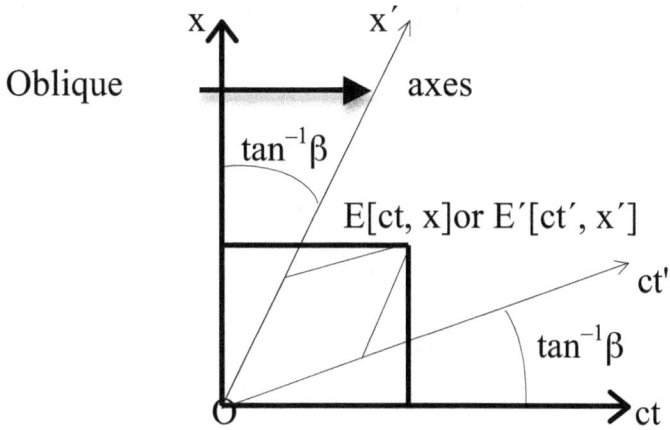

The geometrical form of the Lorentz transformation

The symmetry of space-time means that the transformed axes rotate through equal angles, $\tan^{-1}\beta$. The relativity of simultaneity is clearly exhibited on this diagram: two events that occur at the same time in the ct, x -frame necessarily occur at different times in the oblique ct′, x′-frame.

6. RELATIVISTIC COLLISIONS AND THE CONSERVATION OF 4-MOMENTUM

Consider the interaction between two particles, 1 and 2, to form two particles, 3 and 4. The contravariant 4-momenta are P_i^μ :

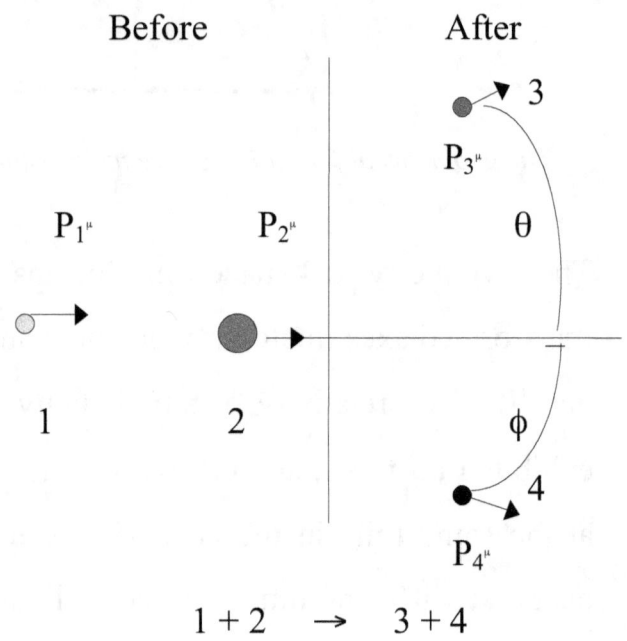

$1 + 2 \rightarrow 3 + 4$

All experiments are consistent with the fact that the 4-momentum of the system is conserved. We have, for the contravariant 4-momentum vectors of the interacting particles,

$$P_1^\mu + P_2^\mu = P_3^\mu + P_4^\mu$$

$\uparrow$ initial free state $\uparrow$ final free state

and a similar equation for the covariant 4-momentum vectors,

$$P_{1\mu} + P_{2\mu} = P_{3\mu} + P_{4\mu}.$$

If we are interested in the change $P_1^\mu \to P_3^\mu$, then we require

$$P_1^\mu - P_3^\mu = P_4^\mu - P_2^\mu$$

and

$$P_{1\mu} - P_{3\mu} = P_{4\mu} - P_{2\mu}.$$

Forming the invariant scalar products, and using $P_{i\mu}P_i^\mu = (E_i^0/c)^2$ we obtain

$$(E_1^0/c)^2 - 2(E_1 E_3/c^2 - \mathbf{p}_1 \cdot \mathbf{p}_3) + (E_3^0/c)^2$$
$$= (E_4^0/c)^2 - 2(E_2 E_4/c^2 - \mathbf{p}_2 \cdot \mathbf{p}_4) + (E_2^0/c)^2$$

Introducing the scattering angles, θ and ϕ, this equation becomes

$$E_1^{0\,2} - 2(E_1 E_3 - c^2 p_1 p_3 \cos\theta) + E_3^{0\,2} =$$
$$E_2^{0\,2} - 2(E_2 E_4 - c^2 p_2 p_4 \cos\phi) + E_4^{0\,2}.$$

If we choose a reference frame in which particle 2 is at rest (the LAB frame), then $\mathbf{p}_2 = 0$ and $E_2 = E_2^0$ so that

$$E_1^{0\,2} - 2(E_1 E_3 - c^2 p_1 p_3 \cos\theta) + E_3^{0\,2} = E_2^{0\,2} - 2E_2^0 E_4 + E_4^{0\,2}$$

The total energy of the system is conserved, therefore

$$E_1 + E_2 = E_3 + E_4 = E_1 + E_2^0$$

or

$$E_4 = E_1 + E_2^0 - E_3$$

Eliminating E_4 from the above "scalar product" equation gives

$$E_1^{0\,2} - 2(E_1 E_3 - c^2 p_1 p_3 \cos\theta) + E_3^{0\,2} = E_4^{0\,2} - E_2^{0\,2} - 2E_2^0(E_1 - E_3)$$

This is the basic equation for interactions in which two relativistic entities in the initial state interact to give two relativistic entities in the final state. It applies equally well to interactions that involve massive and massless entities.

7. RELATIVISTIC INELASTIC COLLISIONS

We shall consider an inelastic collision between a particle 1 and a particle 2 (initially at rest) to form a composite particle 3. In such a collision, the 4-momentum is conserved (as it is in an elastic collision), however, the kinetic energy is not conserved. Part of the kinetic energy of particle 1 is transformed into excitation energy of the composite particle 3. This excitation energy can take many forms — heat energy, rotational energy, and the excitation of quantum states at the microscopic level. The inelastic collision is as shown.

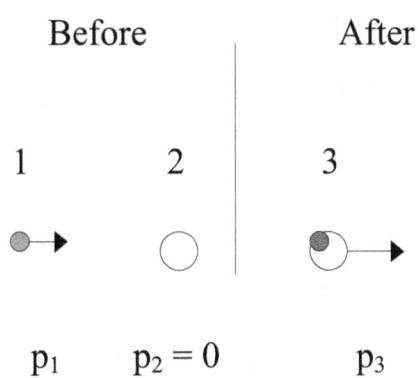

Rest energy: E_1^0 E_2^0 E_3^0
Total energy: E_1 $E_2 = E_2^0$ E_3
3-momentum: p_1 $p_2 = 0$ p_3
Kinetic energy: T_1 $T_2 = 0$ T_3

In this problem, we shall use the energy-momentum invariants associated with each particle, directly:

i) $E_1^2 - (p_1c)^2 = E_1^{0\,2}$

ii) $E_2^2 \quad = E_2^{0\,2}$

iii) $E_3^2 - (p_3c)^2 = E_3^{0\,2}$.

The total energy is conserved, therefore
$$E_1 + E_2 = E_3 = E_1 + E_2^0.$$

Introducing the kinetic energies of the particles, we have
$$(T_1 + E_1^0) + E_2^0 = T_3 + E_3^0.$$

The 3-momentum is conserved, therefore
$$p_1 + 0 = p_3.$$

Using

$$E_3^{0\,2} = E_3^2 - (p_3c)^2,$$

we obtain

$$E_3^{0\,2} = (E_1 + E_2^0)^2 - (p_3c)^2$$
$$= E_1^2 + 2E_1E_2^0 + E_2^{0\,2} - (p_1c)^2$$
$$= E_1^{0\,2} + 2E_1E_2^0 + E_2^{0\,2}$$
$$= E_1^{0\,2} + E_2^{0\,2} + 2(T_1 + E_1^0)E_2^0 \text{ or}$$
$$E_3^{0\,2} = (E_1^0 + E_2^0)^2 + 2T_1E_2^0$$
$$(E_3^0 > E_1^0 + E_2^0).$$

Using $T_1 = \gamma_1 E_1^0 - E_1^0$, where $\gamma_1 = (1 - \beta_1^2)^{-1/2}$ and $\beta_1 = v_1/c$, we have

$$E_3^{0\,2} = E_1^{0\,2} + E_2^{0\,2} + 2\gamma_1 E_1^0 E_2^0.$$

If two identical particles make a completely inelastic collision then

$$E_3^{0\,2} = 2(\gamma_1 + 1)E_1^{0\,2}.$$

8. THE MANDELSTAM VARIABLES

In discussions of relativistic interactions it is useful to introduce Lorentz invariants known as Mandelstam variables. They are, for the special case of two particles in the initial and final states $(1 + 2 \rightarrow 3 + 4)$:

$s = (P_{1^\mu} + P_{2^\mu})[P_{1_\mu} + P_{2_\mu}]$, the total 4-momentum invariant

$= ((E_1 + E_2)/c, (p_1 + p_2))[(E_1 + E_2)/c, -(p_1 + p_2)]$

$= (E_1 + E_2)^2/c^2 - (p_1 + p_2)^2 \rightarrow$ Lorentz invariant

$t = (P_{1^\mu} - P_{3^\mu})[P_{1_\mu} - P_{3_\mu}]$,

the 4-momentum transfer $(1 \rightarrow 3)$ invariant

$= (E_1 - E_3)^2/c^2 - (p_1 - p_3)^2 \rightarrow$ Lorentz invariant,

and

$u = (P_{1^\mu} - P_{4^\mu})[P_{1_\mu} - P_{4_\mu}]$,

the 4-momentum transfer $(1 \rightarrow 4)$ invariant

$= (E_1 - E_4)^2/c^2 - (p_1 - p_4)^2 \rightarrow$ Lorentz invariant

Now,

$sc^2 = E_1^2 + 2E_1E_2 + E_2^2 - (p_1^2 + 2p_1 \cdot p_2 + p_2^2)c^2$

$= E_1^{0\,2} + E_2^{0\,2} + 2E_1E_2 - 2p_1 \cdot p_2 c^2$

$$= E_1^{0\,2} + E_2^{0\,2} + 2(E_1, p_1 c)[E_2, -p_2 c].$$

Lorentz invariant

The Mandelstam variable sc^2 has the same value in all inertial frames. We therefore evaluate it in the LAB frame, defined by the vectors

$$[E_1^L, p_1^L c] \text{ and } [E_2^L = E_2^0, -p_2^L c = 0],$$

so that

$$2(E_1^L E_2^L - p_1^L \cdot p_2^L c^2) = 2E_1^L E_2^0,$$

and

$$sc^2 = E_1^{0\,2} + E_2^{0\,2} + 2E_1^L E_2^0.$$

We can evaluate sc^2 in the center-of-mass (CM) frame, defined by the condition

$$p_1^{CM} + p_2^{CM} = 0$$

(the total 3-momentum is zero):

$$sc^2 = (E_1^{CM} + E_2^{CM})^2.$$

This is the square of the total CM energy of the system.

The Total CM Energy and Production of New Particles

The quantity $c\sqrt{s}$ is the energy available for the production of new particles, or for exciting the internal structure of particles. We can now obtain the relation between the total CM (center-of-mass) energy and the LAB energy of the incident particle (1) and the target (2), as follows:

$$sc^2 = E_1^{0\,2} + E_2^{0\,2} + 2E_1^L E_2^0 = (E_1^{CM} + E_2^{CM})^2 = W^2, \text{ say.}$$

Here, we have evaluated the left-hand side in the LAB frame, and the right-hand side in the CM frame. At very high energies, $c\sqrt{s} \gg E_1^0$ and E_2^0, the rest energies of the particles in the initial state, in which case

$$W^2 = sc^2 \approx 2E_2^L E_2^0.$$

The total CM energy W available for the production of new particles therefore depends on the square root of the incident laboratory energy. This result led to the development of colliding, or intersecting, beams of particles (such as protons and anti-protons) in order to produce sufficient energy to generate particles with rest masses > 100 times the rest mass of the proton (~ 10^9 eV).

9. THE FORMAL STRUCTURE OF LORENTZ TRANSFORMATIONS

The square of the invariant interval s, between the origin of a spacetime coordinate system and an arbitrary event $x^\mu = [x^0, x^1, x^2, x^3]$ is, in index notation

$$s^2 = x^\mu x_\mu = x'^\mu x'_\mu, \text{ (sum over } \mu = 0, 1, 2, 3).$$

The lower indices can be raised using the metric tensor

$$\eta_{\mu\nu} = \text{diag}(1, -1, -1, -1),$$

so that

$$s^2 = \eta_{\mu\nu} x^\mu x^\nu = \eta_{\mu\nu} x'^\mu x'^\nu,$$

(sum over μ and ν).

The vectors now have contravariant forms.

In matrix notation, the invariant is

$$s^2 = \mathbf{x}^T \boldsymbol{\eta} \mathbf{x} = \mathbf{x}'^T \boldsymbol{\eta} \mathbf{x}'.$$

(The transpose must be written explicitly).

The primed and unprimed column matrices (contravariant vectors) are related by the Lorentz matrix operator, **L**

$$\mathbf{x}' = \mathbf{L}\mathbf{x}.$$

We therefore have

$$\mathbf{x}^T \boldsymbol{\eta} \mathbf{x} = (\mathbf{L}\mathbf{x})^T \boldsymbol{\eta} (\mathbf{L}\mathbf{x})$$
$$= \mathbf{x}^T \mathbf{L}^T \boldsymbol{\eta} \mathbf{L} \mathbf{x}.$$

The **x**'s are arbitrary, therefore

$$\mathbf{L}^T \boldsymbol{\eta} \mathbf{L} = \boldsymbol{\eta}.$$

This is the *defining* property of the Lorentz transformations.

The set of all Lorentz transformations is the set *L* of all 4 × 4 matrices that satisfies the defining property

$$L = \{\mathbf{L}: \mathbf{L}^T \boldsymbol{\eta} \mathbf{L} = \boldsymbol{\eta}; \mathbf{L}: \text{all } 4 \times 4 \text{ real matrices};$$
$$\boldsymbol{\eta} = \text{diag}(1, -1, -1, -1)\}.$$

(Note that each **L** has 16 (independent) real matrix elements, and therefore belongs to the 16-dimensional space, R^{16}).

10. ELECTROMAGNETISM: RELATIVITY IN ACTION

The source of the electric field is that attribute of matter called electric charge. Certain naturally occurring materials possess another attribute that generates a field called the magnetic field that has been studied in detail. Whereas electric charges have been isolated and studied, no magnetic charges have ever been observed. The question therefore arises: "what is the source of the magnetic field?" An important clue to solving this mystery was first given by Oersted in the early 1800's. He discovered that a conducting wire that carries an electric current generates a magnetic field in the region around the wire. It is the flow of current (motion of charges) that is somehow related to the magnetic field. We are therefore led to discuss the following ideas: we know that when dealing with motion of any kind, it is essential to take into account the Principle of Relativity, and the space-time transformations of Lorentz and Einstein. To discuss the nature of a force, we must deal with questions concerning mass and acceleration. These two quantities are always measured in a definite frame of reference. In the case of acceleration, we are concerned with measurements of space and time. Previously, we have found that such measurements involve the Lorentz transformation of space-time coordinates between inertial frames. We therefore expect that any discussion of electric and magnetic forces must be made within the framework of Lorentz transformations.

The electrostatic field associated with a charge +Q (the "source charge") at rest in a frame of reference modifies the properties of the space around it in a way that influences a second electric charge +q in the vicinity. The charge +q experiences a force, **F**, due to the stationary charge +Q given by $|\mathbf{F}| = kqQ/r^2$ where k is a constant, and r is the distance between the centers of the charges. The electric field, E(r), at r is $E(r) = F(r)/q$, the force per unit charge at r.

The question therefore arises: "how does the force F(r) change when the source of the field +Q is observed to be in motion?".

Consider two parallel (to the x'–axis) conducting plates with an electric field established between them at rest in an inertial frame, S'. A charge +q' with a mass m' moves with constant acceleration a' in the – y' direction under the influence of the field, $E' = V'/d'$, where d' is the distance between the plates. The frame S' moves to the right at a constant speed V_x with respect an inertial frame S. In S', the constant electric field E' depends on the total charge +Q' on a plate, and on A', the area of the plate:

$$E' = KQ'/A', \text{ where K is a}$$

constant.

In S',

$$F' = E'q' \text{ (Q' is at rest)}$$

$$= m'a'.$$

The key question is:

"Does $E' = E$, the field measured by observers in S?"

We construct a table of the observations made in S and S', and relate them using the standard (Lorentz) transformations of Special Relativity.

	OBSERVERS in S	OBSERVERS in S'	
Charge:	$+q$	$+q'$	$q = q'$

(charge is a Lorentz-invariant)

Mass:	m	m'	$m = \gamma m'$

$$\gamma = \{1 - (V_x/c)^2\}^{-1/2}$$

Distance:	Δy	$\Delta y'$	$\Delta y = \Delta y'$

(y-motion perpendicular to V_x)

Time:	Δt	$\Delta t'$	$\Delta t = \gamma \Delta t'$
Acceleration:	a	a'	
	$F = ma$	$F' = m'a'$	
	$E = F/q$?	$E' = F'/q'$	

(Source $+Q'$ at rest in S')

Plate area $A = L_x \cdot L_z$ $A' = L'_{x'} \cdot L'_{z'}$ $L'_{x'} = \gamma L_x$

$L'_{z'} = L_z$

(perpendicular to V_x)

We can now relate E to E' as follows:

using the listed transformations, we have

$$E' = KQ'/L'_{x'} \cdot L'_{z'}$$
$$= KQ/\gamma L_x \cdot L_z \quad (Q = Q')$$
$$= KQ/\gamma A$$
$$= E/\gamma$$

The electric fields, measured it S' and S are not equal. Since $\gamma > 1$, E' (the field in the rest frame of the plates) is less than E. This is readily understood: the total charge on the plates is unchanged by the motion and therefore the field of the moving plates must increase because the total charge spreads out over a smaller area (L_x is contracted).

We now make use of the fact that $\Delta y' = \Delta y$ (perpendicular to V_x).

The standard result for distance moved under constant acceleration, a, gives

$a(\Delta t)^2/2 = a'(\Delta t')^2/2$, where $a' = F'/m' = E'q'/m'$. Substituting the above transformations in this equation, and rearranging, gives
$$a = (E'q'/m')(1/\gamma) = q(E/\gamma)(\gamma/m)(1/\gamma^2)$$
therefore,
$$F = ma = Eq/\gamma^2 \neq Eq.$$
The force on the charge q, according to observers in S, involves the velocity-dependent term $1/\gamma^2$.

We have
$$F = (1 - \beta^2)Eq, \text{ where } \beta = V_x/c$$
$$= Eq - (V_x^2/c^2)Eq$$
$$= Eq - (V_x E/c^2)V_x q$$

The term $(V_x E/c^2)$ is called the magnetic field, B_{mag}.

Therefore,
$$F = E_{elec}q - B_{mag}V_x q$$

Force is a vector quantity and therefore this equation must, in general, be written in vector form:
$$\mathbf{F} = \mathbf{E}_{elec}q + q\mathbf{V}_x \times \mathbf{B}_{mag} \text{ (+ sign by convention)}$$

The Lorentz equation for the electromagnetic force is therefore

$$\mathbf{F}_{EM} = q\mathbf{E} + q\mathbf{V}_x \times \mathbf{B} = \mathbf{F}_E + \mathbf{F}_M$$

where $\mathbf{V}_x \times \mathbf{B}$ is a vector product.

The motion of charged particles in a magnetic field is determined by the Lorentz force.

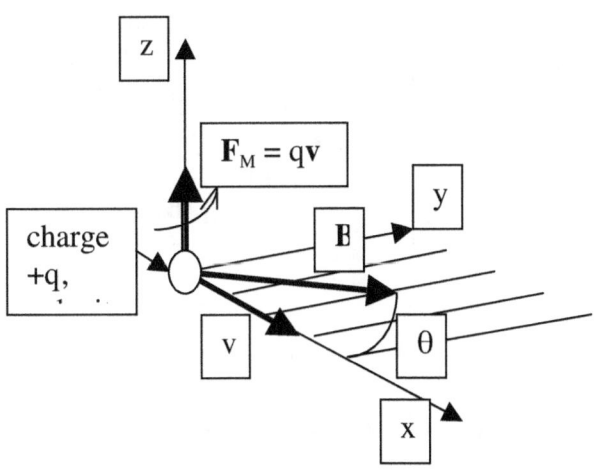

Modern, circular, charged-particle accelerators use the Lorentz equation as a starting point in their design.

If the angle θ is 90° then the three vectors **v**, **B**, and **F**$_M$ are mutually orthogonal. This means that, if **v** is perpendicular to **B**, a charge +q will move in a circular path when placed in a uniform magnetic field **B**:

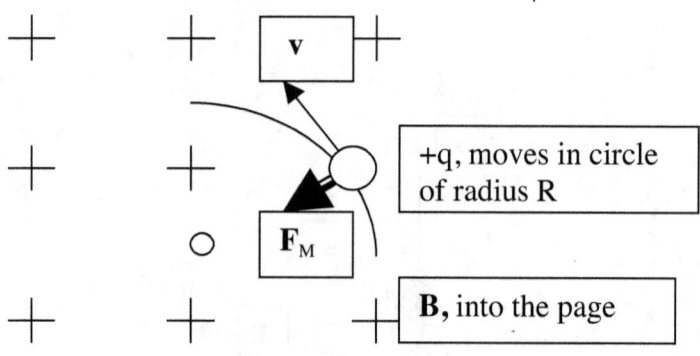

The radius of the circular path of q is given by R = p/qB, where **p** is the (relativistic) momentum of the particle.

We see that the magnetic component of the field is an entirely relativistic effect; no "magnetic charges" are involved.

11. A RELATIVISTIC THEORY OF FIELDS

The fundamental experiments of Oersted, Faraday and Henry carried out in the early part of the 19th century, were put into a unified mathematical theory of electromagnetism by Maxwell in 1865. We shall see that his theory is fully consistent with Special Relativity and provides the basis for future developments of all field theories. His original paper contained no fewer than twenty differential equations that made it difficult to interpret the full meaning of the theory. By chance, a new and powerful mathematical method − Vector Analysis − was introduced in the early 1880's by Gibbs in the US and Heaviside in the UK. Heaviside was keenly interested in Maxwell's theory and he immediately interpreted the many Maxwell equations using his method of Vector Analysis. He reduced the original set of Maxwell equations to the following four vector equations that remain the standard form, to this day; they are (SI units):

$\nabla \cdot \mathbf{E} = \rho/e_0 = \text{div } \mathbf{E}$ (Gauss' Law)

$\nabla \times \mathbf{B} = (1/c^2)\partial \mathbf{E}/\partial t + (1/c^2 e_0)\mathbf{j} = \mathbf{curl\ B}$

(Ampere - Maxwell Law)

$\nabla \cdot \mathbf{B} = 0$ (no magnetic charges)

$\nabla \times \mathbf{E} = -\partial \mathbf{B}/\partial t$ (Faraday's Law)

    E is the electric field vector

    B is the magnetic field vector

    ρ is the charge density

    j is the current density

and e_0 is the permittivity of the vacuum.

(Later, we shall introduce μ_0, the permeability of the vacuum).

The Ampere – Maxwell equation is consistent with charge conservation:

div **j** = div $(-e_0 \partial \mathbf{E}/\partial t)$ (note div **curl B** = 0)

    = $-\partial\rho/\partial t$

or

    div **j** + $\partial\rho/\partial t = 0$, the equation of charge conservation.

Electromagnetic Waves

We consider a vacuum state in which there are no free charges or currents. The Ampere-Maxwell equation then contains no current density term.

Taking the **curl** of Faraday's Law we have

curlcurl E $+ \partial/\partial t(\text{curl } \mathbf{B}) = 0$.

Now,

curlcurl = **grad**div $-\nabla^2$

and

curl B $= (1/c^2)\partial E/\partial t$ (no current density **j** in the Ampere-Maxwell Law)

therefore

graddiv **E** $- \nabla^2 \mathbf{E} + \partial/\partial t((1/c^2)\, \partial/\partial t(\mathbf{E})) = 0$.

But div **E** = 0 therefore

$\nabla^2 \mathbf{E} - (1/c^2)\, \partial^2 E/\partial t^2 = 0$.

This is a vector wave equation in which **E** propagates at the invariant speed of light, c. The above argument can be used to obtain a similar equation for the propagation of the **B** – field.

Before Maxwell derived his set of equations, it was known that the product of the measured values of the permittivity μ_0 and permeability e_0 was directly related to the measured value of the speed of light, c:

$$\mu_0 e_0 = 1/c^2.$$

Maxwell used this fact in developing his general theory, and was thereby led to assert the *equivalence of electromagnetism and light waves.*

In Electrostatics, Coulombs Law can be used to develop the concept of the *potential* at a point in space, ϕ_a, as follows:

The work done by the electrostatic field of a charge +Q, located at the origin of the x-axis, in moving a charge +q from a point a to a point h along the x-axis is

$$\begin{aligned}
W_{ab} &= \int_{a,b} dW \\
&= \int_{a,b} F_x . dx \\
&= \int_{a,b} kQq/x^2\, dx \\
&= kQq[-1/x]_{a,b} \\
&= q\{kQ/a - kQ/b\} \\
&= q\{\phi_a - \phi_b\} \text{ where } \phi_a \text{ is the potential at a}
\end{aligned}$$

due to Q.

It is usual to define the (arbitrary) zero of potential at b = ∞, then the potential at a is ϕ_a = kQ/a.

The solution of Maxwell's equation

$$\nabla \cdot \mathbf{B} = 0$$

is that **B** is the curl of a vector, **A** (say) the *3 – vector potential*:

$$\mathbf{B} = \text{curl}\mathbf{A}.$$

Using the Maxwell equation

$$\nabla \times \mathbf{E} = - \partial \mathbf{B}/t = - \partial(\nabla \times \mathbf{A})/\partial t$$
$$= - \nabla \times \partial \mathbf{A}/\partial t,$$
$$\nabla \times (\mathbf{E} + \partial \mathbf{A}/\partial t) = 0$$
$$= \nabla \times (\text{grad of the scalar potential, } \phi), \text{ therefore}$$
$$\mathbf{E} = \nabla \phi - \partial \mathbf{A}/\partial t \text{ (minus sign is by convention)}.$$

We have obtained solutions of two of the Maxwell equations.

In the formal theory of Special Relativity, we have introduced *Lorentz invariants* associated with 4 - dimensional space - time: they include

the 4 - velocity $V_\mu V^\mu$ and the 4 - momentum $P_\mu P^\mu$. In all cases, the first component of the associated 4 - vector is a *scalar* and the remaining three components form a *3 - vector*. In developing a relativistic theory of Electromagnetism we therefore introduce 1) a 4 - potential A^m with components [ϕ, c**A**] in which ϕ is the scalar potential and **A** is the 3 - vector potential and 2) a 4 - current density J^μ with components $\rho_0\gamma$[c, v] = [cρ, **j**] where ρ is the charge density and **j** is the 3 – current density.
The 4 - dimensional differential operators are:

$\partial_\mu = [\partial/c\partial t, \partial/\partial x, \partial/\partial y, \partial/\partial z]$

and $\partial^\mu = [\partial/c\partial t, -\partial/\partial x, -\partial/\partial y, -\partial/\partial z]$,

and their product is

$\partial_\mu \partial^\mu = (1/c^2)\partial^2/\partial t^2 - \nabla^2$

$= \square$ the d'Alembertian.

We have

$\partial_\mu J^\mu = \partial\rho/\partial t + \text{div}\mathbf{j} = 0$, the equation of charge conservation, and the d'Alembertian that, when operating on the free - space components of A^μ generates fields that propagate at the speed of light, c.

In Particle Dynamics, the Minkowski 4 – force is defined as

$$F^\mu = dP^\mu/dt = d(m_0 V^\mu)/dt =$$

$$d[\gamma m_0 c, m_0 \mathbf{v}]/dt = d[E/c, \mathbf{p}]/dt$$

where m_0 is the rest mass, **v** is the relativistic velocity, E the total energy and **p** is the relativistic momentum.

Differentiating $E^2 - (pc)^2 = E^{02}$ with respect to t, we obtain

$$F^\mu = [\mathbf{v}_N \cdot \mathbf{f}/c, \mathbf{f}] = [f^0, f^k] \text{ (where } \mathbf{f} = d\mathbf{p}/dt\text{)}.$$

The Lorentz 3 – force $\mathbf{F}_L$ is related to the **E** and **B** fields acting on a particle of charge q that is moving with a velocity $\mathbf{v}_N$ as follows:

$$\mathbf{F}_L = q(\mathbf{E} + \mathbf{v}_N \times \mathbf{B}).$$

In 4 – dimensions we therefore expect F_μ to be related to the 4 – velocity of the particle as follows

$$F_\mu = (q/c) F_{\mu\nu} V^\nu \text{ (sum over } \nu)$$

where

$$F_{\nu\mu} = \begin{matrix} 0 & E^x & E^y & E^z \\ -E^x & 0 & -cB^z & cB^y \\ -E^y & cB^z & 0 & -cB^x \\ -E^z & -cB^y & cB^x & 0 \end{matrix}$$

is the covariant anti – symmetric field tensor, and $V^\nu = [\gamma c, \mathbf{v}]$ is the 4 – velocity.

In F_μ, the sum runs over a repeated index $\nu = 0, 1, 2, 3$. For example,

if $\mu = 0$ then

$$F_0 = (q/c) \{F_{00} V^0 + F_{01} V^1 + F_{02} V^2 + F_{03} V^3\}$$
$$= (q/c) \{ 0 + E^x v^x + E^y v^y + E^z v^z \}$$
$$= (q/c) \{ \mathbf{E} \bullet \mathbf{v} \} \text{ where } \mathbf{v} = \gamma \mathbf{v_N}.$$

If $\mu = 1$ then

$$F_1 = (q/c) \{-E^x \gamma c + 0 + -cB^z v^y + cB^y v^z\}$$

or $-F_1 = \gamma q \{ \mathbf{E} + \mathbf{v_N} \times \mathbf{B} \}_{\text{x-component}}$

This procedure can be repeated for all the components.

We therefore see that the 4 – dimensional form F_μ provides a concise statement of the Lorentz force and also leads to the relation

$$F_0 = (q/c)\, \mathbf{E} \cdot \mathbf{v}.$$

12. RELATIVITY AND PARTICLE – WAVE DUALITY

Particle-Wave Duality is one of the most profound concepts in Physics. Originally, deBroglie introduced the concept using a geometrical model. He realized, however, that he needed to base such a novel and highly controversial concept on a firmer basis, namely, the fundamental invariants of Special Relativity.

In his geometrical approach, deBroglie supposed that the quantum wave associated with an electron orbiting a nucleus forms a stationary pattern (a standing wave) that fits exactly into an orbit of given radius, R_n i.e.

$$n\lambda_n = 2\pi R_n,$$

where n is an integer and λ_n is the wavelength of the standing wave in the nth-orbit. Such a standing wave represents a *stationary charge distribution*, and therefore *no radiation occurs*.

If the quantum wave in an orbit does *not* have a wavelength that obeys the above condition, then the waves destructively interfere and *no wave exists*. If there is no wave then there is *no particle.* The essential arguments in deBroglie's formal theory are as follows:

We recall that, in general, the displacement of a wave when at a radial distance **r** from a chosen origin, at a definite time, t, has the form:

$$\psi(\mathbf{r},t) = A\cos(\omega t - \mathbf{k}\cdot\mathbf{r}),$$

where $\omega = 2\pi\upsilon$ is the angular frequency, **r** is the radius vector, and $|\mathbf{k}| = 2\pi/\lambda$ is the wave number. *deBroglie recognized that the phase*

$$[(\omega/c)(ct) - \mathbf{k}\cdot\mathbf{r}]$$

is a Lorentz invariant. (For mathematical convenience, we have introduced the invariant speed of light, c, to give "time" units of distance). The explicit form of the invariant phase is

$$K^{\mu T}E_\mu = \text{Lorentz invariant},$$

where $K^\mu = [\omega/c, \mathbf{k}]$ is the frequency-wave number column four vector, and $E_\mu = [ct, -\mathbf{r}]$ is the event column four vector. The lower index on E indicates that we have chosen **r** to be negative. This is a mathematical formality. We can appreciate deBroglie's new concept by considering the following argument: events [ct, **r**] form the basis of all discussions of the dynamics of particles and the propagation of waves. Beginning with the event E^μ[ct, **r**] we can follow two routes to systems of particles or waves:

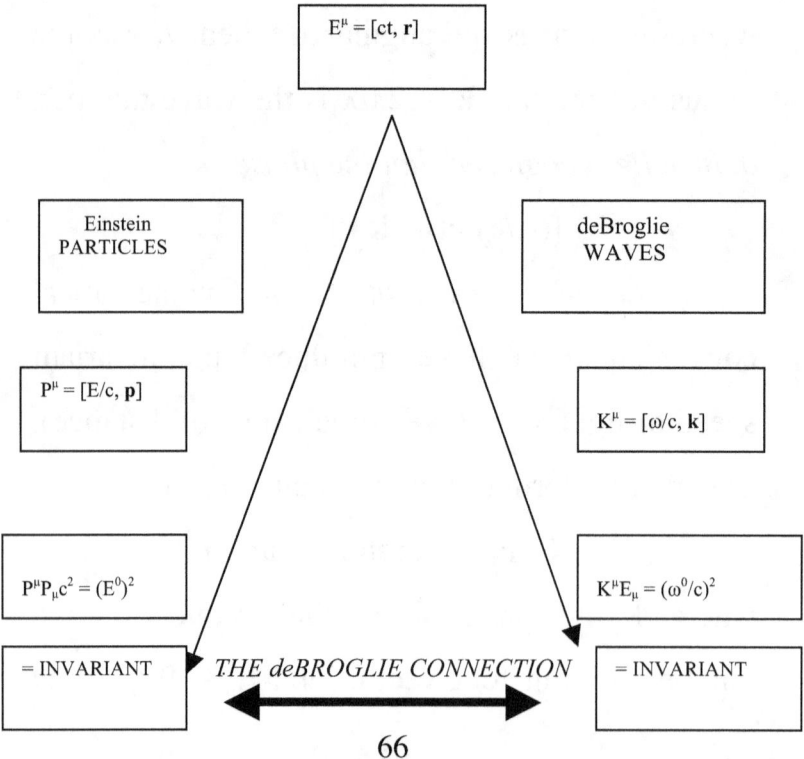

Here, E^0 is the invariant rest energy of a particle, and ω^0 is the invariant (angular) frequency of the wave. P^μ is the energy-momentum four-vector. According to deBroglie, the Lorentz four-vectors P^μ and K^μ, and their Lorentz invariants, are directly linked through a common space-time event four-vector E^μ. He surmised that the connection should be as simple as possible, namely, they are proportional to each other:

P^μ = constant x K^μ

To agree with the Planck relation for a quantum of electromagnetic radiation, $E = h\upsilon$, where E is the energy and υ the frequency of the radiation, he chose the constant to be $h/2\pi$, where h is Planck's constant. *In complete generality, he therefore proposed the equation*

$$P^\mu = (h/2\pi)K^\mu.$$

Equating the components of the four vectors gives two equations

$$E/c = (h/2\pi)(\omega/c) = (h/2\pi)(2\pi\upsilon/c) = h\upsilon/c,$$

and therefore

$$E = h\upsilon,$$

and

$$|\mathbf{p}| = (h/2\pi)|\mathbf{k}| = (h/2\pi)(2\pi/\lambda),$$

so that

$$p = h/\lambda.$$

These two relations link our concepts of a particle and a wave: the particle energy is proportional to the frequency of the equivalent wave, and the particle momentum is proportional to its equivalent inverse wavelength.

In our everyday experiences, we do not encounter this duality because the value of Planck's constant is so small and therefore the wavelengths of everyday objects are immeasurably small. The value of the constant is $h = 6.6... \times 10^{-34}$ Joule-seconds.

At the atomic level, however, the duality is a standard feature. For example, an electron has a mass of about 10^{-30} kilograms. If it moves with a speed $v = 10^6$ meters/second then its wavelength is

$\lambda = h/mv = 6.6 \times 10^{-34}/10^{-30} \times 10^6 = 6.6 \times 10^{-10}$ meters, a value comparable with the size of an atom.

13. RELATIVITY AND deBROGLIE MATTER-WAVES

The angular frequency ω and the wave vector $\mathbf{k}$ of a wave, moving in free space, are the components of a four-vector – the frequency-wave number column four-vector

$$K^\mu = [\omega/c, \mathbf{k}] \text{ (a contravariant vector)}$$

in which $\omega = 2\pi\upsilon$, the angular frequency, $|\mathbf{k}| = 2\pi/\lambda$ where λ is the wavelength, c is the invariant speed of light, and $\mu = 0, 1, 2, 3$.
$K_\mu = [\omega/c, -\mathbf{k}]$ is the related covariant vector.
The "scalar product" of these vectors has the Lorentz-invariant form

$$K^{\mu T} K_\mu = (\omega/c)^2 - k^2$$
$$= (\omega^0/c)^2 \text{ an invariant in all}$$

inertial frames of reference.
Rearranging, we obtain

$$\omega^2 = k^2 c^2 + \omega^{0\,2}$$

or

$$\omega = c[k^2 + (\omega^0/c)^2]^{1/2}.$$

This relation between ω and k is called a *dispersion law*.

Note that, if $\omega^0 = 0$ (true for electromagnetic waves) then

$$\omega = kc$$

or

$$2\pi\upsilon = 2\pi c/\lambda$$

and therefore,

$$c = \upsilon\lambda, \text{ the familiar form.}$$

We wish to find the dispersion law for deBroglie matter-waves. The fundamental deBroglie relation between the four-momentum P^μ [E/c, **p**] and the four-vector K^μ is

$$P^\mu = \hbar K^\mu, \text{ where } \hbar = h/2\pi, \text{ the}$$

normalized Planck's constant.

In component form, we have

$$E = \hbar\omega \text{ and } \mathbf{p} = \hbar\mathbf{k}$$

where E is the total energy and **p** is the relativistic 3-momentum.

The rest energy of a particle is given by

$$E^0 = \hbar\omega^0$$

which leads to

$$m^0 c^2 = \hbar\omega^0,$$

in which m^0 is the rest mass of the particle, and therefore

$$(\omega^0)^2 = (m^0 c^2/\hbar)^2 = (E^0/\hbar)^2.$$

The dispersion law for the corresponding matter-wave is then

$$\omega = c[k^2 + (m^0 c/\hbar)^2]^{1/2} = c(k^2 + K^{0\,2})^{1/2}$$

where

$$K^0 = (m^0 c/\hbar) = (E^0/\hbar c) = \text{a constant for a given particle.}$$

The phase velocity of the matter-wave is defined in the conventional way

$$v_\varphi = \omega/k = c[1 + (K^0/k)^2].$$

For a particle with a finite rest mass, $K^0 > 0$ in which case the phase velocity is greater than c. We therefore conclude that the phase velocity of a matter-wave cannot be identified with the velocity of the corresponding particle. This would seem to be an insurmountable obstacle for the theory. However, we know from experience that classical waves can range over large distances whereas classical particles are "localized". We may require "localized waves" even in the everyday case; these can be achieved by forming "wave packets". A wave packet is a superposition of many waves of slightly differing wavelength and phase.

The simplest case involves the superposition of two waves with slightly different frequencies and wave numbers: $\omega \pm \Delta\omega$ and $k \pm \Delta k$, respectively. The resultant amplitude, Y, as a function of time and space is, for waves of unit amplitude:

$$Y = \exp\{i[(k + \Delta k)x - (\omega + \Delta\omega)t]\}$$
$$+ \exp\{i[(k - \Delta k)x - (\omega - \Delta\omega)t]\}.$$

Rearranging, and substituting a cosine for the difference of exponentials, gives

$$Y = \exp\{i(kx - \omega t)\}\cos(\Delta k x - \Delta\omega t),$$

a single wave " modulated" by the cosine term.

The "phase" velocity of the single wave is $v_\varphi = \omega/k$, and the velocity of the modulation envelope is

$$v_G = \Delta\omega/\Delta k,$$ called the "group velocity" of the resultant waveform. In the limit, as $\Delta k \to 0$,

$$v_G = d\omega/dk = (d/dk)[c(k^2 + K^{0\,2})^{1/2}]$$
$$= kc(k^2 + K^{0\,2})^{-1/2}$$
$$= kc^2/\omega.$$

We have seen that $v_\varphi = \omega/k$, therefore, we obtain *the fundamental deBroglie invariant*

$$v_\varphi v_G = c^2,$$ the wave equivalent of Einstein's equation $E/m = c^2$.

But,

$$p = \hbar k \text{ and } E = \hbar\omega,$$

therefore,\
$$v_G = (p/\hbar)(\hbar/E)c^2 = pc^2/E = \gamma m^0 vc^2/\gamma m^0 c^2$$
$$= v, \text{ the velocity of the particle.}$$

(The relativistic mass of the particle is given by γm^0).

We can therefore associate, in a one-to-one correspondence, the motion of a deBroglie wave packet and the equivalent particle. de Broglie's profound discovery led directly to a formal theory of Quantum Mechanics, including the Uncertainty Principle.

14. ENERGY AND MOMENTUM OPERATORS IN QUANTUM MECHANICS: THE SCHROEDINGER EQUATION

A plane harmonic wave of unit amplitude can be represented by the complex form

$$\psi(t, r) = \exp\{-i(\omega t - \mathbf{k}\cdot\mathbf{r})\}$$

where

$\mathbf{r} = [x, y, z]$, the radial coordinate,

$\omega = 2\pi\upsilon$, the angular frequency,

and

$|\mathbf{k}| = 2\pi/\lambda$, the wave number.

The trigonometric (real) versions of $\psi(t, \mathbf{r})$ can be obtained by taking either the real or the imaginary part of the complex function.

To describe the motion of the wave moving through time and space we require the rates-of-change of ψ with respect to t and $\mathbf{r}$.

If we limit the discussion to the x-component of the wave-motion, the time-derivative of $\psi(t, x)$ is

$$\partial\psi(t, x)/\partial t = -i\omega\exp\{-i(\omega t - k_x x)\},$$
$$= -i\omega\psi(t, x), (\omega \text{ and } k_x \text{ are constants})$$

and, therefore

$$i\partial\psi(t, x)/\partial t = \omega\psi(t, x).$$

The differential operator $i\partial/\partial t$, operating on ψ, multiplies ψ by ω; $i\partial/\partial t$ is the equivalent of ω. The x-derivative of ψ is

$$\partial\psi(t, x)/\partial x = ik_x\exp\{-i(\omega t - k_x x)\}$$
$$= ik_x\psi(t, x),$$

or

$$-i\partial\psi(t, x)/\partial x = k_x\psi(t, x).$$

The differential operator $-i\partial/\partial x$ operating on ψ, multiplies ψ by k_x; $-i\partial/\partial x$ is the equivalent of k_x.

We are interested in the application of the above discussion of *classical wave motion* to the motion of a *quantum mechanical deBroglie matter-wave.*

The fundamental deBroglie particle-wave equations are
$$E = h\omega/2\pi \text{ and } p = hk/2\pi.$$
The angular frequency ω, and the wave number k_x are interpreted in terms of their equivalent operators
$$\omega = 2\pi E/h = i\partial/\partial t$$
or
$$E = (ih/2\pi)\partial/\partial t, \text{ the } \textit{energy operator},$$
and
$$k_x = 2\pi p_x/h = -i\partial/\partial x,$$
or
$$p_x = (-ih/2\pi)\partial/\partial x, \text{ the } \textit{momentum operator}.$$

In the absence of external forces, a non-relativistic particle of mass m, moving with a velocity v, has a kinetic energy, E, given by
$$E = mv^2/2 = p^2/2m$$

A quantum mechanical description of the particle, moving in one dimension, can be obtained by making the *substitutions*

$$E \to (ih/2\pi)\partial/\partial t,$$

and $\quad p_x \to (-ih/2\pi)\partial/\partial x,$

leading to the operator equation

$$(-\hbar^2/2m)\partial^2/\partial x^2 = (i\hbar)\partial/\partial t$$

where $\hbar = h/2\pi$.

These operators must operate on something, in this case, it is the deBroglie wave function, $\Psi(t, x)$, that "represents" the particle. We therefore obtain

$$(-\hbar^2/2m)\partial^2\Psi(t, x)/\partial x^2 = i\hbar\partial\Psi(t, x)/\partial t = E\Psi(t, x)$$

This is the Schroedinger equation of a particle of mass m with kinetic energy E, moving in free space.

Localized particles, moving under the influence of forces, are represented by wave-packets — superpositions of deBroglie matter-waves. *The concept of energy and momentum operators continues to hold in such cases; each component of the wave-packet then has different values of energy and momentum.*

If a particle is moving under the influence of a force the Schroedinger equation becomes

$(-\hbar^2/2m)\partial^2\Psi(t, x)/\partial x^2 + V(t, x)\Psi(t, x)$
$= i\hbar\partial\Psi(t, x)/\partial t.$

The inclusion of the potential V is *axiomatic*; the original motivation for including such a term was based on the classical form of the total energy of the system, namely, the Hamiltonian H:

H = T + V, where T is the kinetic energy.

The time-dependent Schroedinger equation holds if the particle is not free, and if it is not in a definite state of energy.

If the potential is *time-independent*

(V(t, x) – > V(x)) we have

$(-\hbar^2/2m)\partial^2\Psi(t, x)/\partial x^2 + V(x)\Psi(t, x) = E\Psi(t, x),$

where the state has a definite energy, E.

To solve this equation, the wave function

Ψ(t, x) is written as the *product of two functions*, one dependent on time, and the other dependent on position, thus

$$\Psi(t, x) = \phi(t)\psi(x).$$

Here, $\psi(x)$ is the time-independent wave function that satisfies the eigenvalue equation

$$H_S\psi(x) = E\psi(x), \text{ (E is an eigenvalue of } H_s)$$

where

$$H_s = (-\hbar^2/2m)\partial^2/\partial x^2 + V(x)$$

is the Schroedinger operator. The complete solution of the equation requires the imposition of boundary conditions.

15. THE FOURIER – RAYLEIGH – deBROGLIE – HEISENBERG UNCERTAINTY PRINCIPLE

In the late 19th century, the renowned physicist, Lord Rayleigh, derived a fundamental constraint for the characteristic properties of "wave packets", formed by the superposition of waves of differing wavelengths. The constraint applies not only to pulses of electromagnetic waves used to transmit information around the globe, and to distant satellites, but also to quantum wave packets that consist of underlying deBroglie matter waves.

Lord Rayleigh's argument can be understood by considering the simplest case of the superposition of two sine-waves with wavelengths
λ_1 and λ_2 where $\lambda_1 = 2\lambda_2/3$. The sum of the two waves generates a basic wave packet with its "wavelength" λ:

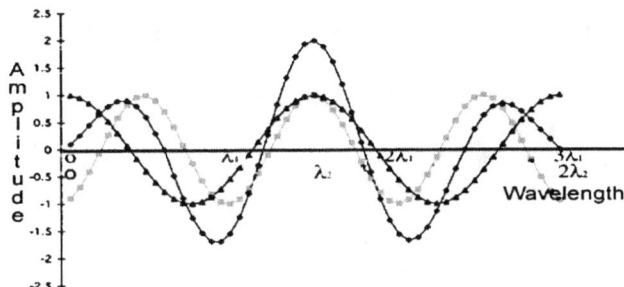

<- Δx ->

Let the extent of the wave packet be Δx; in this special case,

$$\Delta x = 3\lambda_1,$$

and,

$$\Delta x = 2\lambda_2$$

In general, we see that if

$$\Delta x = n\lambda_2, \text{ where n is an integer}$$

then,

$$\Delta x \geq (n + 1)\lambda_1$$

(This is an algebraic statement of the fact that there must be at least one more wavelength, λ_1 to generate the zeros at the extremes of Δx).

Subtracting the equations we obtain
$$\Delta x/\lambda_1 - \Delta x/\lambda_2 \geq (n + 1) - n,$$
therefore
$$\Delta x[(\lambda_2 - \lambda_1)/\lambda_1\lambda_2] \geq 1.$$
Put
$$\lambda_2 = \lambda_1 + \Delta\lambda_1,$$
so that
$$\Delta x[\Delta\lambda_1/(\lambda_1^2 + \lambda_1\Delta\lambda_1)] \geq 1.$$
But $\lambda_1^2 \gg \lambda_1\Delta\lambda_1$ because we can choose $\Delta\lambda_1$ to be arbitrarily small, therefore
$$\Delta x \Delta\lambda_1/\lambda_1^2 \geq 1.$$
We note, however, that in terms of differentials,
$$|d(1/\lambda)| = |d\lambda/\lambda^2|,$$
therefore putting $\lambda_1 \approx \lambda$, the "wavelength" of the wave packet, we have
$$\Delta x \Delta(1/\lambda) \geq 1.$$
This is a general result, true for all forms of wave packets.

Lord Rayleigh expressed this constraint in the form

$$\Delta t \Delta \upsilon \geq 1,$$

where Δt is the duration of the wave packet (or pulse), and $\Delta \upsilon$ is the frequency range associated with the wave packet. For example, if $\Delta t = 10^{-6}$ seconds (1μsec), a "fast" electromagnetic pulse in the field of electronics 50 years ago, then $\Delta \upsilon \geq 10^6$ cycles per second (1MHz). If $\Delta t = 10^{-9}$ seconds (1nanosec), a typical "fast" pulse in contemporary electronics, then $\Delta \upsilon \geq 10^9$ cycles per second (1GHz). In this form, the constraint is of great importance in modern communications.

In 1924, deBroglie proposed the fundamental equation of particle-wave duality:

p (the particle momentum) = h/λ (the particle "wavelength"), where h is Planck's constant. We can therefore write Rayleigh's constraint for a wave packet formed by the superposition of deBroglie waves as

$$\Delta x \Delta(1/\lambda) \to \Delta x \Delta(p/h) \geq 1,$$

therefore,

$$\Delta x \Delta p \geq h.$$

This is Heisenberg's Uncertainty Principle, published in 1926. (A more exact limit is $\Delta x \Delta p \geq h/4\pi$). Heisenberg's famous principle is a direct consequence of the Rayleigh constraint and the concept of particle-wave duality, introduced by deBroglie. (Heisenberg developed his principle in terms of a theory of measurement). We shall see that it is fitting to do justice to the pioneers in this field by referring to the fundamental principle as the

Fourier-Rayleigh-deBroglie-Heisenberg Uncertainty Principle.

Uncertainty Principles – Classical and Quantum

The mathematical origin of all "uncertainty principles" – classical and quantum – is the relationship between Fourier transform pairs. In the time (t) – angular frequency (ω) domain, the Fourier transform pair (suitably normalized) is:

$$f(t) = (1/\sqrt{(2\pi)}) \int_{-\infty,\infty} F(\omega) \exp\{-i\omega t\} d\omega$$

and

$$F(\omega) = (1/\sqrt{(2\pi)}) \int_{-\infty,\infty} f(t) \exp\{i\omega t\} dt.$$

It is of interest to consider the functions f to be probability density functions in general, and a normal Gaussian distribution in particular:

$$f(t) = (1/\sigma\sqrt{(2\pi)}) \exp\{-(t-\mu)^2/2\sigma^2\}$$

with Fourier transform

$$F(\omega) = (1/2\pi\sigma) \int_{-\infty,\infty} \exp\{-(t-\mu)^2/2\sigma^2 + i\omega t\} dt.$$
$$= (1/2\pi\sigma) \int_{-\infty,\infty} \exp\{-at^2 + bt - c\} dt,$$

where

$$a = 1/2\sigma^2, b = \mu/\sigma^2 + i\omega, \text{ and } c = \mu^2/2\sigma^2.$$

The integral in the Fourier transform of the Gaussian distribution, $F(\omega)$, can be evaluated by developing the standard integral

$$I = \int_{-\infty, \infty} \exp\{-x^2\}\,dx = \sqrt{\pi}$$

as follows:

let the exponent x^2 be replaced by the quadratic form

$$x^2 \rightarrow ax^2 - bx + c$$

to give the integral

$$I' = \int_{-\infty, \infty} \exp\{-ax^2 + bx - c\}\,dx.$$

Introduce a change of variable

$$u = x\sqrt{a} - b/2\sqrt{a} \text{ so that } du/\sqrt{a} = dx$$

then

$$\int_{-\infty, \infty} \exp\{-ax^2 + bx - c\}\,dx$$
$$= (1/\sqrt{a})\exp\{(b^2 - 4ac)/4a\} \int_{-\infty, \infty} \exp\{-u^2\}\,du$$
$$= (\sqrt{\pi/a}) \exp\{(b^2 - 4ac)/4a\}.$$

The Fourier transform, $F(\omega)$, is therefore

$$F(\omega) = \exp\{\mu^2/(2\sigma^2)\}[\exp\{-(\omega - i\mu/\sigma^2)^2/(2/\sigma^2)\}]/[\sqrt{(2\pi)}/\sigma].$$

If f has zero mean ($\mu = 0$) then

$$F(\omega) = [1/(\sqrt{(2\pi)}/\sigma]\exp\{-\omega^2/(2/\sigma^2)\}.$$

In this special case, the Fourier transform of the Gaussian distribution f(t), with zero mean and standard deviation σ, is also a Gaussian distribution with zero mean but with a standard deviation equal to (1/ σ). The product of their standard deviations is therefore unity, and in terms of their variances:

var(f(t))•var(F(ω)) = 1

This property is a limiting case of a general inequality associated with the product of the variances of Fourier transform pairs. A detailed analysis of the problem shows that, if f is an arbitrary probability density distribution, and F is its Fourier transform, then:

var(f)•var(F) ≥ 1.

We see that it is not possible to concentrate, arbitrarily, both f(t) and F(ω) – the more f(t) is "squeezed in", the more F(ω) is "spread out". This result is a classical "uncertainty principle".

In Quantum Mechanics, the operators **p** and **q** that represent generalized momentum and position coordinates (in the Hamilitonian sense), obey the commutation relation

$$qp - pq = ih/2\pi$$

where h is Planck's constant.

The relation between the (conjugate) variables **p** and **q** can be written in terms of the relation between Fourier transform pairs.

In the Schroedinger representation the observable **q** is a diagonal operator, and the momentum operator is defined as

$$p = -(ih/2\pi)d/dq.$$

In the Dirac representation, **p** is taken as a diagonal operator.

Let $<q|S>$ and $<p|S>$ be the probability amplitudes that measurements corresponding to the operators **p** and **q** give the eigenvalues p and q, then we find

$$<p|S> = (2\pi/h) \int_{-\infty,\infty} <q|S> \exp\{-2\pi iqp/h\}dq$$

and

$$<q|S> = (2\pi/h) \int_{-\infty,\infty} <p|S> \exp\{2\pi iqp/h\}dp,$$

for a state **S**.

The probability amplitude distributions of two (conjugate) variables are the Fourier transforms of each other. Following the general argument presented above, we conclude that the variances of the probability amplitude distributions of conjugate variables in Quantum Mechanics satisfy some such inequality. Heisenberg's Uncertainty Principle for conjugate pairs of observables therefore follows directly from the fact that the observables are Fourier transforms of each other. The critical concept that permits the application of the theory to Quantum Mechanics is deBroglie's renowned particle – wave duality, summarized by the equations:

$E = h\omega/2\pi$ and $p = hk/2\pi = h/\lambda$

where E is the energy, p is the momentum, $\omega = 2\pi\nu$ where ν is the frequency, $k = 2\pi/\lambda$ where λ is the wavelength, and h is Planck's constant.

Now, frequency ν is directly related to inverse time t, and wavelength has units of distance, therefore

$$E \cdot t = h = p \cdot \lambda$$

energy and time, and momentum and distance, are conjugate variables. We therefore expect that the product of the variances of any two conjugate quantum variables obey an inequality limited by Planck's quantum of action, h. This argument is the basis of Heisenberg's Uncertainty Principle.

16. THE KLEIN – GORDON EQUATION

An important development took place in the mid – 1920's when the energy – momentum invariant of Special Relativity was combined with the energy and momentum operators from deBroglie's particle – wave duality of matter, as follows:

The invariant of relativistic dynamics is
$$E^2 - (\mathbf{p}c)^2 = E^{o2}.$$
If the substitutions
$E \rightarrow i\hbar \partial/\partial t$, $p^x \rightarrow -i\hbar \partial/\partial x$, $p^y \rightarrow -i\hbar \partial/\partial y$, and $p^z \rightarrow -i\hbar \partial/\partial z$ are made then
$$(i\hbar \partial/\partial t)^2 = (-i\hbar c)^2 \{(\partial/\partial x)^2 + (\partial/\partial y)^2 + (\partial/\partial z)^2\} = E^{o2},$$
therefore
$$-\hbar^2 \partial^2/\partial t^2 + \hbar^2 c^2 \nabla^2 = E^{o2}$$
where
$$\nabla^2 = \partial^2/\partial x^2 + \partial^2/\partial y^2 + \partial^2/\partial z^2.$$
The left hand side of this equation is an operator – it operates on the wave function of a particle $\psi(t, \mathbf{r})$, giving
$$\nabla^2 \psi(t, \mathbf{r}) - (1/c^2) \partial^2 \psi(t, \mathbf{r})/\partial t^2 = (E^o/\hbar c)^2 \psi(t, \mathbf{r}).$$

This is the Klein – Gordon relativistic wave equation of particles with spin – zero (or integer – spin). Typical of such particles are *mesons* – the carriers of the nuclear force between nucleons.

17. THE DIRAC EQUATION

The energy – momentum invariant of a freely moving particle with rest energy E^0, total energy E and relativistic momentum **p** is

$$E^2 - (\mathbf{p}c)^2 = E^{02}$$

so that the total energy is

$$E = +/- \{(\mathbf{p}c)^2 + E^{02}\}.$$

In Classical Mechanics, a negative total energy of a freely moving particle is without meaning because kinetic energy is necessarily a positive quantity. However, in Quantum Mechanics the negative sign associated with E is found to play a key role in the description of the creation and annihilation of matter and anti-matter; these topics will be treated in a later section.

We have met the quantum transformations $E \rightarrow i\hbar\partial/\partial t$ and $\mathbf{p} \rightarrow -i\hbar\nabla$ Substituting these transforms in Einstein's expression for the energy, we obtain

$$i\hbar\partial/\partial t = +/- \{-\hbar^2 c^2 \nabla^2 + E^{02}\}^{1/2}$$

This is an operator equation – it operates on a wave function, ψ, that describes the quantum mechanical particle of rest energy E^0. Dirac was concerned with the space-time symmetry of this operator equation; the left-hand side is linear in time whereas the right-hand side involves the space variables squared. The right-hand side cannot be linearized simply by taking the square root because of the presence of the rest-energy term, E^{02}. He set out to make the right-hand side a perfect square and put

$$(pc)^2 + E^{02} = (p^xc)^2 + (p^yc)^2 + (p^zc)^2 + E^{02}$$

$$= (ca_xp^x + ca_yp^y + ca_zp^z + bE^0)^2$$

$$= c^2a_x^2p^{x2} + c^2a_xp^xa_yp^y + c^2a_xp^xa_zp^z + ca_xp^xbE^0$$

$$\qquad + c^2a_y^2p^{y2}$$

$$\qquad\qquad + c^2a_z^2p^{z2}$$

$$\qquad\qquad\qquad + b^2E^{02}$$

This equation is true if

$$a_x^2 = a_y^2 = a_z^2 = b^2 = I,$$

and all the cross-terms vanish in pairs:

$$a_x a_y + a_y a_x = 0 \text{ etc.}$$

We see that the a's and b must anti-commute:

$$a_x a_y = -a_y a_x \quad a_x a_z = -a_z a_x \ \ldots\ a_x b = -b a_x \text{ etc.}$$

The quantities a_i and b are to be regarded as operators that operate on the wave function ψ of the particle, just as the differential operators do.

The total energy E is equivalent to the Hamiltonian H of the particle and therefore

$$H = E = \{(\mathbf{p}c)^2 + E^{02}\}^{1/2}$$

$$= c\mathbf{a} \cdot \mathbf{p} + bE^0$$

where $\mathbf{a} = [a_x,\ a_y,\ a_z]$

Following Dirac, we write the quantities a_i and b using Greek symbols:

$$\mathbf{a} \rightarrow \boldsymbol{\alpha} = [\alpha_x,\ \alpha_y,\ \alpha_z] \text{ and } b \rightarrow \beta$$

The Dirac Hamiltonian can be written in terms of the differential operators:

$$H_D = -ci\hbar\boldsymbol{\alpha}\cdot\nabla + \beta E^0.$$

The wave equation of the free particle is therefore

$$H_D\psi = E\psi.$$

The forms of the Dirac matrices and the wave functions ψ are discussed in the next section.

18. DIRAC MATRICES, WAVEFUNCTIONS, ELECTRONS AND POSITRONS

Dirac showed that the matrices α_i and β that appear in his Hamiltonian are *4 x 4 matrices*. He had a choice of representations of the matrices that obey the anti-commutation rules; his chosen set includes generators, σ_i, of the group SU(2):

$$\alpha_x = \begin{pmatrix} 0 & 0 & & \\ & & \sigma_x & \\ 0 & 0 & & \\ & & 0 & 0 \\ \sigma_x & & & \\ & & 0 & 0 \end{pmatrix}$$

$$\alpha_y = \begin{pmatrix} 0 & 0 & & \\ & & \sigma_y & \\ 0 & 0 & & \\ & & 0 & 0 \\ \sigma_y & & & \\ & & 0 & 0 \end{pmatrix}$$

$$\alpha_z = \begin{pmatrix} \sigma_z & 0 & 0 \\ 0 & 0 & \\ 0 & 0 & \\ & & \sigma_z \\ & 0 & 0 \\ & 0 & 0 \end{pmatrix}$$

$$\beta = \begin{pmatrix} I & & \\ 0 & 0 & \\ 0 & 0 & \\ & & -I \\ & 0 & 0 \\ & 0 & 0 \end{pmatrix}$$

where

$$\sigma_x = \begin{pmatrix} 0 & 1 \\ 1 & 0 \end{pmatrix}, \quad \sigma_y = \begin{pmatrix} 0 & -i \\ i & 0 \end{pmatrix}, \quad \sigma_z = \begin{pmatrix} 1 & 0 \\ 0 & -1 \end{pmatrix}$$

are the *Pauli spin matrices* that belong to SU(2).

These matrices were introduced by Pauli to explain the observed splitting of quantum states in atoms by Stern and Gerlach; they are associated with the two states of the *intrinsic spin* of the electron.

The 4 x 4 matrix operators α and β must operate on 4 – component quantities. The wave function ψ is therefore a 4 – component (column) quantity called a "spinor". Dirac introduced the concept of *particles and anti-particles* – in the first case, the electron and its anti-particle, the positron. The 4 components of ψ then correspond to the electron with "spin up" and "spin down" and the positron with "spin up" and "spin down".

19. INTRODUCTION TO LIE GROUPS

Galois, shortly before his death in a duel at the age of twentyone, showed that an equation of degree ≥ 5 cannot, in general, be solved by algebraic means. In the course of this great work, he developed the ideas of Lagrange, Abel, and Ruffini, and introduced the concept of a *group*. Consider the roots

of the equation $x^4 = 1$ (the fourth roots of unity); they are $\{x\} = \{1, i, -1, -i\}$ where $i = \sqrt{-1}$. These roots have the properties:

1. The product of any two of them is a member of the set $\{x\}$. *The set obeys closure.*

2. The elements of the set $\{x\}$ obey the *associative operation.*

3. There exists in the set $\{x\}$ a unique *identity element* (in this example it is 1) such that the product of the identity element and any element is equal to the element itself.

4. The set {x} contains *inverse elements* such that the product of any element and its inverse is equal to the identity.

Note that, in the fourth roots of unity, the inverse of each element is its reciprocal.

The set {x} in which the elements obey these four rules is said to form a group: $\{x\} = \{g_1 = e, g_2, g_3, \ldots g_n\} = G_n$ – a *group of order n* with identity element e.

Cayley, who introduced the algebra of matrices in the mid-nineteenth century, first recognized that the elements of a group can be numbers, matrices, operators, etc.

Groups can be formed in which the law of composition is *addition* and not *multiplication*. For example, the integers {…−3, −2, −1, 0, 1, 2, 3…} form a group under *addition*; in this case, the identity is zero, and the inverse of 3 is −3, etc.

20. THE SPECIAL UNITARY GROUP, SU(2)

In Modern Physics, Lie groups play a key role. The Lorentz transformations form a Lie group, and the Pauli spin matrices are the generators of a Lie group.

The special unitary group in 2-dimensions, SU(2), is defined as the set of 2 x 2 matrices with the properties:

SU(2) = {$UU^\dagger = 1$, det$U = +1$, u_{ij} ε complex numbers}

where

$$U = \begin{pmatrix} a & b \\ c & d \end{pmatrix} \quad U^\dagger = \begin{pmatrix} a^* & c^* \\ b^* & d^* \end{pmatrix}$$

The defining conditions mean that U must have the form

$$U = \begin{pmatrix} a & b \\ -b^* & a^* \end{pmatrix}$$

The infinitesimal form of U is

$$U_{inf} = \begin{pmatrix} 1 + \delta a & \delta b \\ -\delta b^* & 1 + \delta a^* \end{pmatrix}$$

The determinantal condition is

$$(1 + \delta a)(1 + \delta a^*) + \delta b \delta b^* = 1,$$

so that to 1st-order,

$$1 + \delta a^* + \delta a = 1, \text{ or } \delta a = -\delta a^*,$$

therefore

$$U_{inf} = \begin{pmatrix} 1 + \delta a & \delta b \\ -\delta b^* & 1 - \delta a \end{pmatrix}$$

Introducing, explicitly, real and imaginary components of the matrix elements, gives

$$U_{inf} = \begin{pmatrix} 1 + i\delta\alpha/2 & (\delta\beta + i\delta\gamma)/2 \\ (-\delta\beta + i\delta\gamma)/2 & 1 - i\delta\alpha/2 \end{pmatrix}$$

(The factor of ½ has been included for later convenience).

Now, any 2 x 2 matrix can be written as the following linear combination

$$\begin{pmatrix} a & b \\ c & d \end{pmatrix} = [(a+d)/2]\begin{pmatrix} 1 & 0 \\ 0 & 1 \end{pmatrix} + [(b+c)/2]\begin{pmatrix} 0 & 1 \\ 1 & 0 \end{pmatrix} + [i(b-c)/2]\begin{pmatrix} 0 & -i \\ i & 0 \end{pmatrix} + [(a-d)/2]\begin{pmatrix} 1 & 0 \\ 0 & -1 \end{pmatrix}$$

The infinitesimal form of U is therefore

$$U_{inf} = \begin{pmatrix} 1 & 0 \\ 0 & 1 \end{pmatrix} + (i\delta\gamma/2)\begin{pmatrix} 0 & 1 \\ 1 & 0 \end{pmatrix} + (i\delta\beta/2)\begin{pmatrix} 0 & -i \\ i & 0 \end{pmatrix} + (i\delta\alpha/2)\begin{pmatrix} 1 & 0 \\ 0 & -1 \end{pmatrix}.$$

The matrices

$$\sigma_x = \begin{pmatrix} 0 & 1 \\ 1 & 0 \end{pmatrix} \quad \sigma_y = \begin{pmatrix} 0 & -i \\ i & 0 \end{pmatrix} \text{ and}$$

$$\sigma_z = \begin{pmatrix} 1 & 0 \\ 0 & -1 \end{pmatrix}$$

are the *Pauli spin matrices*.

The expression for U_{inf} is seen to have the standard Lie form $U_{inf} = I + i\Sigma_{j=1 \text{ to } 3} (\sigma_j/2)\delta\alpha_j$. I and $\sigma[\sigma_x, \sigma_y, \sigma_z]$ form a complete set of 2 x 2 Hermitian matrices.

21. LIE'S CONTINUOUS TRANSFORMATION GROUPS

Sophus Lie was concerned with finding general solutions of differential equations and not of linear equations. In the following discussion, the subject will be introduced from a geometrical and not from an algebraic point of view. Rotations of Cartesian coordinates through an angle φ (a continuous variable), can be described in terms of a rotation matrix operator, $\mathbf{R}_C(\varphi)$:

From the diagram, we see that

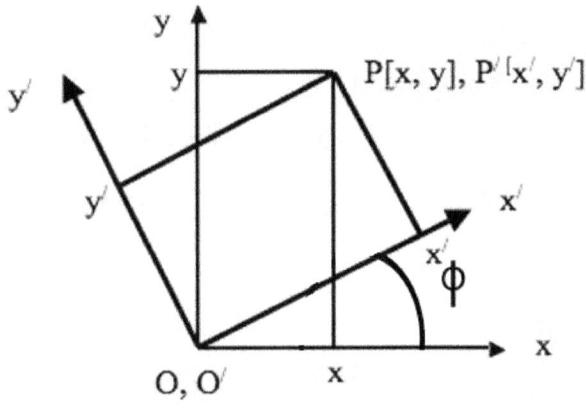

$$x' = x\cos\varphi + y\sin\varphi$$
$$y' = -x\sin\varphi + y\cos\varphi$$

or, in matrix form

$$\begin{pmatrix} x' \\ y' \end{pmatrix} = \begin{pmatrix} \cos\varphi & \sin\varphi \\ -\sin\varphi & \cos\varphi \end{pmatrix} \begin{pmatrix} x \\ y \end{pmatrix}$$

This can be written as an operator equation

$$\mathbf{P'} = \mathbf{R}_C(\varphi)\,\mathbf{P}$$

which means that the coordinate rotation operator $\mathbf{R}_c(\varphi)$ transforms the column vector $\mathbf{P}[x, y]$ to the column vector $\mathbf{P'}[x', y']$.

The properties of $\mathbf{R}_C(\varphi)$ lead naturally to the notion of a Lie Group. A rotation through a small angle $\delta\varphi$ can be written

$$\mathbf{R}_C(\delta\varphi) = \begin{pmatrix} 1 - \delta\varphi^2/2! \ldots & \delta\varphi - \delta\varphi^3/3! \ldots \\ -\delta\varphi + \delta\varphi^3/3! \ldots & 1 - \delta\varphi^2/2! \ldots \end{pmatrix}$$

If the Maclaurin expansions are limited to 1st-order in $\delta\varphi$ then

$$\mathbf{R}_C(\delta\varphi) = \begin{pmatrix} 1 & \delta\varphi \\ -\delta\varphi & 1 \end{pmatrix}$$

108

therefore,
$$x' = x + \delta\varphi \cdot y$$
$$y' = -\delta\varphi \cdot x + y$$

The infinitesimals δx and δy are
$$\delta x = x' - x = y\delta\varphi$$
and
$$\delta y = y' - y = -x\delta\varphi.$$

Lie recognized that these expressions can be written in terms of *differential operators*, as follows
$$\delta x = (y\partial/\partial x - x\partial/\partial y)x\delta\varphi$$
and
$$\delta y = (y\partial/\partial x - x\partial/\partial y)y\delta\varphi$$

Putting $x = x_1$ and $y = x_2$, we can write
$$\delta x_i = \mathbf{X}x_i\delta\varphi, \; i = 1, 2$$
where
$$\mathbf{X} = (x_2\partial/\partial x_1 - x_1\partial/\partial x_2).$$

The connection with differential equations is made through the operator **X**.

The operator $\mathbf{R}_C(\delta\varphi)$ can be written

$$\begin{pmatrix} 1 & 0 \\ 0 & 1 \end{pmatrix} \begin{pmatrix} 0 & -\delta\phi \\ \delta\phi & 0 \end{pmatrix}$$

or $\quad \mathbf{R}_c(\delta\varphi) = \mathbf{I} - \mathbf{i}\delta\varphi$

where

$\quad\quad\quad\quad$ **I** is the identity operator

and

$$\mathbf{i} = \begin{pmatrix} 0 & -1 \\ 1 & 0 \end{pmatrix} \quad (\mathbf{i}^2 = -\mathbf{I}).$$

Performing two infinitesimal rotations in succession gives

$\mathbf{R}_C^2(\delta\varphi) = (\mathbf{I} - \mathbf{i}\delta\varphi)(\mathbf{I} - \mathbf{i}\delta\varphi)$

$\quad\quad\quad = \mathbf{I} - 2\mathbf{i}\delta\varphi$, to 1st – order in $\delta\varphi$,

$\quad\quad\quad = \mathbf{R}_C(2\delta\varphi).$

If n successive rotations through $\delta\varphi$ are carried out, then

$\mathbf{R}_C^n(\delta\varphi) = \sum_{r=0 \text{ to } n} \{n!/(n-r)!r!\}\{(-\mathbf{i}\delta\varphi)^r\},$

$\quad\quad\quad\quad\quad\quad\quad\quad\quad\quad$ (exactly).

This sum can be written in terms of the series expansions of the sine and cosine functions, giving $\mathbf{R}_C^n(\delta\varphi) = \mathbf{I}\cos(n\delta\varphi) - \mathbf{i}\sin(n\delta\varphi)$
$$= \exp\{-\mathbf{i}n\delta\varphi\}$$
$$= \mathbf{R}_C(n\delta\varphi).$$

If, as $n \to \infty$ and $\delta\varphi \to 0$, the product $n\delta\varphi \to \varphi$, a finite limit, then

$\mathbf{R}_C^n(\delta\varphi) = \mathbf{R}_C(\varphi) = \mathbf{I}\cos\varphi - \mathbf{i}\sin\varphi$
$$= \exp\{-\mathbf{i}\varphi\}.$$

A finite rotation can be built up by exponentiation of infinitesimal rotations, each rotation being as near to the identity as possible. This is an essential result of Lie's Theory of Continuous Transformations; it can be relatively straightforward to obtain the infinitesimal form of a transformation when its finite form is intractable.

22. ROTATIONS, ANGULAR MOMENTUM, AND LIE GROUPS

In Classical Mechanics, the angular momentum of a mass m moving in a plane at a distance **r** from the origin with a momentum **p** is

$$\mathbf{L}_Z = \mathbf{r} \times \mathbf{p}.$$

Introducing the coordinates of the mass m, and the components of **p**, we have

$$L_Z = (xp_y - yp_x).$$

The transition from Classical to Quantum Mechanics is made by carrying out the following transformations

$$p_x \rightarrow -i(h/2\pi)\partial/\partial x \text{ and } p_y \rightarrow -i(h/2\pi)\partial/\partial y.$$

The angular momentum, L_Z, is then written in operator form

$$\mathbf{L}_z = -i(h/2\pi)(x\partial/\partial y - y\partial/\partial x).$$

We have introduced the notation

$$\mathbf{X} = (y\partial/\partial x - x\partial/\partial y), \text{ and therefore,}$$

$$\mathbf{L}_Z = i(h/2\pi)\mathbf{X}$$

and

$$\mathbf{X} = -i(2\pi/h)\mathbf{L}_Z.$$

The infinitesimals are then

$$\delta x_i = \mathbf{X} x_i \delta\varphi = -i(2\pi/h)\mathbf{L}_Z x_i \delta\varphi.$$

Lie discussed infinitesimal transformations in the following way:

If the rotation matrix, $\mathbf{R}_C(\varphi)$, is differentiated with respect to φ then

$$d\mathbf{R}_C(\varphi)/d\varphi = \begin{pmatrix} -\sin\phi & \cos\phi \\ -\cos\phi & -\sin\phi \end{pmatrix}$$

This expression is evaluated at the value of φ that gives the identity, namely at $\varphi = 0$, leading to

$$d\mathbf{R}_C(\varphi)/d\varphi \big|_{\varphi=} = \begin{pmatrix} 0 & 1 \\ -1 & 0 \end{pmatrix} = -i.$$

We can therefore write

$$\mathbf{R}_C(\delta\varphi) = \mathbf{I} - i\delta\varphi = \exp\{-i\delta\varphi\}$$

$$= \mathbf{I} + d\mathbf{R}_C(\varphi)/d\varphi \big|_{\varphi=0} \delta\varphi.$$

In three dimensions, this becomes

$$\mathbf{R}_C(\delta\alpha_1, \delta\alpha_2, \delta\alpha_3)$$
$$= \mathbf{I} + \sum_{i=1 \text{ to } 3} \partial \mathbf{R}_C(\alpha_1, \alpha_2, \alpha_3)/\partial\alpha_i \Big|_{\text{all } \alpha_i\text{'s} = 0} \cdot \delta\alpha_i$$

23. POSITRON-ELECTRON ANNIHILATION-IN-FLIGHT

A discussion of the annihilation-in-flight of a relativistic positron and a stationary electron provides a topical example of the use of relativistic conservation laws. This process, in which two photons are spontaneously generated, has been used as a source of nearly monoenergetic high-energy photons for the study of nuclear photo-disintegration since 1960. Using the general result for a $1 + 2 \rightarrow 3 + 4$ interaction, we have:

$$E_1^{0\,2} - 2(E_1 E_3 - c^2 p_1 p_3 \cos\theta) + E_3^{0\,2}$$
$$= .E_4^{0\,2} - E_2^{0\,2} - 2E_2^{0}(E_1 - E_3)$$

This equation provides the basis for an exact calculation of this process. In the present case:

$E_1 = E_{pos}$ (the incident positron energy),

$E_2 = E_e^0$ (the rest energy of the stationary electron),

$E_3 = E_{ph1}$ (the energy of photon 1), and

$E_4 = E_{ph2}$ (the energy of photon 2).

The rest energies of the positron and the electron are equal. The general equation now reads

$$E_e^{02} - 2\{E_{pos}E_{ph1} - cp_{pos}E_{ph1}(\cos\theta)\} + 0$$
$$= 0 - E_e^{02} - 2E_e^0(E_{pos} - E_{ph1})$$

therefore

$$E_{ph1}\{E_{pos} + E_e^0 - [E_{pos}^2 - E_e^{02}]^{1/2}\cos\theta\}$$
$$= (E_{pos} + E_e^0)E_e^0,$$

leading to

$$E_{ph1} = E_e^0/(1 - k\cos\theta)$$

where

$$k = [(E_{pos} - E_e^0)/(E_{pos} + E_e^0)]^{1/2}.$$

The *maximum* energy of photon 1, E_{ph1}^{max}, occurs when $\theta = 0$, corresponding to motion in the forward direction; its energy is

$$E_{ph1}^{max} = E_{oe}/(1 - k).$$

If, for example, the incident total positron energy is 30 MeV, and $E_e^0 = 0.511$ MeV then

$$E_{ph1}^{max} = 0.511/[1-(29.489/30.511)^{1/2}] \text{ MeV}$$
$$= 30.25 \text{ MeV}.$$

The forward-going photon has energy equal to the kinetic energy of the incident positron (T_1 = 30 − 0.511 MeV) plus approximately three-quarters of the total rest energy of the positron-electron pair ($2E_e^0$ = 1.02 MeV). Using the conservation of the total energy of the system, we see that the energy of the backward-going photon is about 0.25 MeV.

The method of positron-electron annihilation-in-flight provides one of the very few ways of generating nearly mono-energetic photons at high energies.

24. NEUTRONS, PROTONS AND THE YUKAWA POTENTIAL

Introduction

Nuclear Physics began with Chadwick's discovery of the neutron in early 1932. Chadwick concluded from an analysis of his data that the neutron mass is slightly greater than the proton mass. In the same year, Cockcroft and Walton carried out the first proton-induced nuclear reaction using a high-voltage generator that produced protons with energies up to 600 keV.

Heisenberg introduced a theoretical model of the neutron-proton interaction in which he treated the neutron and proton as two charge states of a single object, the nucleon. Heisenberg's model included space, spin and charge exchange terms in the nuclear Hamiltonian. Fermi introduced a model of β-decay in which a neutron decays into a proton, a spontaneously generated electron and an anti-neutrino. In this model, a neutron is a fundamental particle and not a proton-electron combination as assumed in the original Heisenberg model. Wigner showed that the nucleon-nucleon force is characterized by a short-range attractive potential and a repulsive potential (core) between the nucleons when they are in very close proximity. Breit et al. analyzed data on proton-proton scattering and showed that the nuclear force is the same between neutron-proton and proton-proton pairs, after correcting for the Coulomb interaction. This result led to the hypothesis of charge independence for nuclear forces.

Enter Yukawa

Yukawa proposed a novel mechanism to describe the neutron-proton interaction. In his model, the interaction between nucleons is mediated by a heavy particle, the quantum of the nuclear force field. Based on the limited data available at the time, he estimated the mass of the field quantum (a boson) to be 200 electron masses. The Yukawa model was the first application of the Klein – Gordon equation to a fundamental problem in Nuclear Physics namely to an understanding of the force between nucleons. The Klein – Gordon equation for particles of *zero rest mass*, and with the correct spin is

$$\nabla^2 \psi = (1/c^2)\, \partial^2 \psi/\partial t^2.$$

This is the classical equation of electromagnetism; it has a static solution

$$\psi(r) = -ke^2/r, \quad k = 9 \times 10^9 \text{ in MKS units.}$$

The solution gives the Coulomb potential for the electromagnetic field. The particle in this case is the photon – it is the *mediator* of the electromagnetic force.

For particles with finite rest mass, and with the correct spin, the complete Klein – Gordon equation must be solved. The static solution is

$$\psi(r) = -(\text{constant}^2/r)\exp[-r/r']$$

where $r' = \hbar/m^0 c^2 = \hbar c/E^0$, the Compton wavelength of a particle with a rest mass m^0.

This solution (that can be verified by direct substitution) gives the Yukawa potential for the (meson) field associated with the nucleon – nucleon force. The *meson* is the *mediator* of the nuclear force (Yukawa, 1935).

25. PROBABILITY AND THE MAXWELL-BOLTZMANN DISTRIBUTION

The concept of numerical probability originated in the study of games of chance. However, in the 18th-century, the concept developed abstractly – specific reference to coins, playing cards, dice etc. became unnecessary. The first use of probability in Physics took place in the mid-to-late 1800's when Clausius, Maxwell, Boltzmann, and Gibbs developed the field of Statistical Mechanics. This triumph of intellectual thought continues to have a profound effect throughout the Physical Sciences, particularly in its modern form of Quantum Statistical Mechanics.

The notion of numerical probability is related to belonging to a class; for example, if the probability is 1/6 that the next throw of a dice will be a "1" then this statement has to do with a class of events that includes the specific event.

An aspect is a symbol denoting a distinct state. If an aspect is possible in a ways, and is not possible in b ways then the probability of the aspect occurring is defined as

$$a/(a + b).$$

No condition favors one aspect over another, and therefore all (a + b) aspects have the same chance of occurring. The total number of aspects is dependent upon the knowledge of the observer. We note

$$[a/(a + b)] + [b/(a + b)] = 1, \text{ a certainty}$$

where the first term is the probability of occurrence, and the second term is the probability of non-occurrence.

The probability of two independent events both occurring is the probability given by the product of the two separate, independent probabilities. The question of the independence of the events being studied frequently presents difficulties in dealing with a particular statistical problem.

On tossing an ideal coin there are two possible aspects – a head "H" or a tail "T". It is axiomatic that the ratio (number of heads/number of tails) for a very large number of tosses is equal to 1. The probability of each event is the accepted value of ½.

If we now consider two unlike coins, labeled 1 and 2, (they are distinguishable) then there are said to be 4 complexions:

Complexion	coin 1	coin 2	symbol	probability
I	H	H	$a_1 a_2$	¼
II	T	T	$b_1 b_2$	¼
III	H	T	$a_1 b_2$	¼
IV	T	H	$b_1 a_2$	¼

Here, a –> head, b –> tail, a_1 –> head for coin 1, etc.

The possible, independent complexions, are represented by forming the product

$$(a_1 + b_1)(a_2 + b_2),$$

and the probability of each complexion is ¼.

The probability of a composite event, such as a_1a_2, is seen to be the product of the component events.

If we consider two similar coins, the complexions III and IV are no longer distinguishable, and therefore the number of complexions is reduced to 3. In this case, the events are represented by the terms of the binomial

$$(a + b)^2 = a^2 + 2ab + b^2$$

The probability of a^2 = probability of b^2 = ¼, and the probability of ab = ½, twice that of a^2 and b^2. The probability is greatest for the state with the same number of heads as tails. Explicitly

State	symbol	weight	probability
both H	a^2	1	¼
both T	b^2	1	¼
H and T or T and H	ab	2	½

In the case of three dissimilar coins, there are 8 complexions:

Complexion	Symbol
I	$a_1a_2a_3$
II	$a_1a_2b_3$
III	$a_1b_2a_3$
IV	$a_1b_2b_3$
V	$b_1a_2a_3$
VI	$b_1a_2b_3$
VII	$b_1b_2a_3$
VIII	$b_1b_2b_3$

The possible complexions are obtained by taking the product

$$(a_1 + b_1)(a_2 + b_2)(a_3 + b_3)$$

All complexions are equally probable; the probability of each one is 1/8.

For three similar coins, there are 4 states:

Statistical state	Symbol	Weight	Probability
all H	a^3	1	1/8
all T	b^3	1	1/8
2H + 1T	a^2b	3	3/8
1H + 2T	ab^2	3	3/8

The symbols are represented by the terms of $(a + b)^3$.

The most likely states are those that have the closest approach to equality between the numbers of heads and tails.

We now consider the case of any number of coins, N.

I) All dissimilar: the possible complexions are represented by the product

$$(a_1 + b_1)(a_2 + b_2)(a_3 + b_3)\ldots(a_N + b_N)$$

II) All identical: the possible combinations are given by the terms in the binomial expansion:

$$(a + b)^N = a^N + a^{N-1}b + [N(N-1)/2!]a^{N-2}b^2 + \ldots b^N$$

$$= a^N + {}_NC_1 a^{N-1}b + {}_NC_2 a^{N-2}b^2 + \ldots b^N$$

where ${}_NC_r$ is the number of combinations of N things taken r at a time.

The statistical states are represented by

$$a^N, a^{N-1}b, a^{N-2}b^2, \ldots b^N$$

where $a^r b^s$ symbolizes r heads and s tails; $r + s = N$.

The weight, or possibility number, is the number of different complexions in a statistical state; it is the number of combinations of N things r at a time:

The Maxwell – Boltzmann Distribution

We shall consider an ideal container at constant temperature that contains a very large number, N, of identical molecules of an ideal gas – a gas that consists of point-like masses with interaction energies between pairs that are negligibly small compared with kinetic energies. It is assumed that there are no external forces acting on the system. In a volume of one cm^3, a typical gas at standard temperature and pressure contains approximately 10^{19} molecules; it is therefore impossible to describe the configuration in terms of the coordinates of each molecule. We replace this unobservable idea with the concept of a complexion or possibility number. Let the coordinates of a molecule lie in x –> x + Δx, y + Δy, and z + Δz (the Δ's are small and finite), and let the complete volume be divided into a finite number of cells, each with a volume $\Delta x \Delta y \Delta z$. A molecule in a particular cell corresponds to a coin with a definite aspect, after a toss. A complexion is defined by the aspects of the molecules – the way in which the molecules are distributed among the cells. If there are c cells and a total of N molecules then the number of complexions in which there are

n_1 molecules in cell 1

n_2 molecules in cell 2

.

n_r molecules in cell r

.

n_c molecules in cell c.

is $N! / \prod_{r=1,c} n_r! = P$, the complexion or possibility number.

For N sufficiently large, we again use Stirling's theorem, and obtain

$$\ln P \approx N \ln N - \sum_{r=1,c} n_r \ln n_r$$

(noting that all the numbers n_r must be sufficiently large).

By assigning a maximum value to P, with the constraint that $\sum_{r=1,c} n_r = N$, we can obtain information on the *form of the distribution of molecules in space* – the equilibrium distribution. In this way, we find that the most probable distribution is uniform – the density is the same, everywhere. However, to obtain a complete description of the system, it is necessary to consider not only the spatial distribution but also the velocity distribution of the molecules; we are dealing with a *dynamical* situation. Let the velocity components of a molecule at (x, y, z) be (v_x, v_y, v_z); we now have a six-dimensional space – the *phase space* of the molecule. An aspect of the molecule is found by stating that the x–coordinate is in x –> x + Δx, ... etc. and the x–component of the velocity is in v_x –> v_x + Δv_x, ... etc. A *phase cell* is $\Delta x \Delta y \Delta z \Delta v_x \Delta v_y \Delta v_z$. If each molecule has a mass m then the momentum components at (x, y, z) are p_x, p_y, p_z ($p_x = mv_x$, etc.) and the hyper-phase cell is $\Delta x \Delta y \Delta z \Delta p_x \Delta p_y \Delta p_z$. (It is interesting to note that the units of $\Delta x \Delta p_x$ are "action"; in Quantum Mechanics, this quantity is equal to Planck's constant, h).

We introduce the postulate: *if the phase cells have the same magnitude, any aspect of a given molecule is as probable as any other.*

The *possibility number* of the distribution in phase cells is
$$P = N! / \prod_{r=1,c} n_r!$$
and the *probability* is
$$W = P / c^N.$$

There are two constraints associated with this problem:

1. The *total number of molecules is fixed*
$$\sum_{r=1,c} n_r = N$$

and

2. At constant temperature, the *total energy is fixed* (there are no external forces)

Let the energies characteristic of each cell be ε_1, ε_2, ... ε_r, ... ε_c then the total energy is

$$E = n_1\varepsilon_1 + n_2\varepsilon_2 + \ldots n_r\varepsilon_r + \ldots n_c\varepsilon_c$$

For an ideal gas, the energies, ε_r, are essentially all kinetic.

Following Boltzmann, we assume that **the equilibrium state is the state of maximum probability.**

The defining equations for this problem are

$$P = N! / \prod_{r=1,c} n_r! \;, \; N = \sum_{r=1,c} n_r \;,$$

and $E = \sum_{r=1,c} n_r\varepsilon_r$.

Let $\omega_r = n_r/N$ – known as the *partition function*.

We have $\sum_{r=1,c} \omega_r = 1$ and $n_r = N\omega_r$.

Using Stirling's theorem gives

$$\ln P = N(\ln N - 1) - \sum n_r (\ln n_r - 1)$$
$$= N(\ln N - 1) - \sum N\omega_r(\ln N + \ln\omega_r - 1)$$
$$= N(\ln N - 1) - N\sum \omega_r(\ln N - 1)$$
$$\quad - N\sum \omega_r \ln\omega_r$$
$$= - N\sum \omega_r \ln\omega_r \;.$$

The three defining equations can therefore be written

$$\ln P = - N\sum \omega_r \ln \omega_r$$

$$N = N\sum \omega_r$$

and

$$E = N\sum \varepsilon_r \omega_r$$

The condition of maximum probability is given by

$$\delta \ln P = -N\sum (1 + \ln \omega_r)\delta \omega_r = 0$$

$$\delta N = N\sum \delta \omega_r = 0$$

and

$$\delta E = N\sum \varepsilon_r \delta \omega_r = 0.$$

We introduce the Lagrange undetermined multipliers, λ and β, leading to

$$\sum (\ln \omega_r + \lambda + \beta \varepsilon_r)\, \delta \omega_r = 0.$$

The variations are arbitrary and therefore

$$\ln \omega_r + \lambda + \beta \varepsilon_r = 0,$$

or

$$\boldsymbol{\omega_r = (1/f)\exp\{-\beta \varepsilon_r\} \text{ where } f \text{ is a constant.}}$$

The partition function, given by this equation, is proportional to the molecules with an energy ε_r. The quantity β (inversely proportional to E/N, the average energy of a molecule) is called the distribution constant. Boltzmann showed that the entropy, S, is related to the probability, W, by the equation

S = k lnW, where k is Boltzmann's constant, from which it follows that $\beta = 1/kT$, where T is the absolute temperature. We therefore find that, for the state of maximum probability,

$$\omega_r \sim \exp\{-\varepsilon_r / kT\}$$

This is the form of the Maxwell-Boltzmann function. We note that

$$f\omega_r = \exp\{-\varepsilon_r / kT\}$$

and

$$f\sum\omega_r = \sum\exp\{-\varepsilon_r / kT\}$$

but $\sum\omega_r = 1$, therefore

$$f = \sum\exp\{-\varepsilon_r / kT\},$$

the sum of the partition function.

www.ingramcontent.com/pod-product-compliance
Lightning Source LLC
Chambersburg PA
CBHW070648220526
45466CB00001B/343